Urban Sustainability

The Urban Sustainability Book Series is a valuable resource for sustainability and urban-related education and research. It offers an inter-disciplinary platform covering all four areas of practice, policy, education, research, and their nexus. The publications in this series are related to critical areas of sustainability, urban studies, planning, and urban geography.

This book series aims to put together cutting-edge research findings linked to the overarching field of urban sustainability. The scope and nature of the topic are broad and interdisciplinary and bring together various associated disciplines from sustainable development, environmental sciences, urbanism, etc. With many advanced research findings in the field, there is a need to put together various discussions and contributions on specific sustainability fields, covering a good range of topics on sustainable development, sustainable urbanism, and urban sustainability. Despite the broad range of issues, we note the importance of practical and policy-oriented directions, extending the literature and directions and pathways towards achieving urban sustainability.

The series will appeal to urbanists, geographers, planners, engineers, architects, governmental authorities, policymakers, researchers of all levels, and to all of those interested in a wide-ranging overview of urban sustainability and its associated fields. The series includes monographs and edited volumes, covering a range of topics under the urban sustainability topic, which can also be used for teaching materials.

Youpeng Lu

Urban Form and Urban Heat Island

Towards a Cool Built Environment

Springer

Youpeng Lu
School of Architecture and Urban Planning
Huazhong University of Science
and Technology
Wuhan, Hubei, China

ISSN 2731-6483 ISSN 2731-6491 (electronic)
Urban Sustainability
ISBN 978-981-96-3540-5 ISBN 978-981-96-3541-2 (eBook)
https://doi.org/10.1007/978-981-96-3541-2

The original submitted manuscript has been translated into English. The translation was done using artificial intelligence. A subsequent revision was performed by the author(s) to further refine the work and to ensure that the translation is appropriate concerning content and scientific correctness. It may, however, read stylistically different from a conventional translation.

This Springer imprint is published by the registered company Springer Nature Singapore Pte Ltd.
The registered company address is: 152 Beach Road, #21-01/04 Gateway East, Singapore 189721, Singapore

Preface

As urbanization accelerates globally, cities face mounting challenges from the intensifying urban heat island (UHI) effect. This phenomenon not only compromises environmental quality but also threatens the comfort and well-being of urban residents. While much research has been conducted on the UHI effect, critical gaps remain—particularly in understanding the intricate relationship between urban form and heat island intensity, the spatial variability of these interactions, and the practical translation of findings into urban planning strategies. This book, *Urban Form and Urban Heat Island: Towards a Cool Built Environment*, was written to address these pressing issues.

This book is intended for scholars, urban planners, policymakers, and students who are invested in creating sustainable and resilient urban environments. It provides a comprehensive framework for understanding the role of urban form in shaping the UHI effect, drawing on cutting-edge research methods, including remote sensing, spatial analysis, and urban data modeling. The findings and recommendations are designed to bridge the gap between academic research and practical urban planning, offering actionable insights for professionals and decision-makers alike.

The book is organized to guide readers through a logical progression, beginning with the theoretical underpinnings of the UHI effect and moving towards empirical analyses and practical solutions. Using Wuhan, a rapidly urbanizing metropolitan area in central China, as a case study, the research identifies key urban form indicators influencing the UHI effect, reveals spatial patterns of heat intensity, and proposes differentiated urban form control strategies. Each chapter builds on the last, culminating in a set of targeted recommendations for mitigating the UHI effect in diverse urban contexts.

The contributors to this volume include experts from various disciplines, whose insights and efforts have been instrumental in shaping the book's narrative. By bringing together perspectives from urban climatology, environmental science, and urban planning, this book aspires to advance the field and inspire new directions for research and practice.

It is our hope that this work will serve as a valuable resource for those striving to create cooler, more livable cities in the face of growing environmental challenges.

Wuhan, China Youpeng Lu

Acknowledgements This work was supported by the National Natural Science Foundation of China (grant numbers 52208088)

I would like to express my sincere gratitude to Prof. Chong Peng from Huazhong University of Science and Technology for her invaluable guidance. Special thanks also go to master's students Wei Zhai, Yuchen Ding, and Chang Ge for their significant contributions to figure drawing and data compilation.

Competing Interests The author has no competing interests to declare that are relevant to the content of this manuscript.

Contents

Abbreviations

UHI	Urban Heat Island
MODIS	Moderate-resolution Imaging Spectroradiometer
UC map	Urban Climate

Chapter 1
Introduction

This chapter aims to elucidate the practical and academic background of this book while defining the focus of the research. Currently, global urbanization faces the challenge of the urban heat island (UHI) effect. The UHI phenomenon refers to the higher temperatures observed in urban areas compared to their surrounding rural areas. In recent years, the process of urbanization has driven the densification and increasing complexity of urban forms, exacerbating the UHI effect. Existing studies have provided effective methods to measure the UHI effect and urban forms, revealing a close relationship between the two. This has opened up possibilities for mitigating the UHI effect through urban planning and management. However, the diversity and complexity of urban forms mean that previous studies have not yet identified the key urban form indicators influencing UHI intensity. In other words, among the many urban form indicators, which ones have the most significant impact on UHI intensity remains unclear.

1.1 Research Background

1.1.1 Global Warming and Urbanization Change Urban Thermal Environment

There is solid evidence that the global climate has been warming over the past century. According to the Fifth Assessment Report of the Intergovernmental Panel on Climate Change (IPCC), the global surface average temperature has been rising steadily from 1880 to 2012. This warming was particularly significant in the 1980s, with the period from 1983 to 2012 being the hottest 30 years in the past 1400 years. Data also show that from 1906 to 2005, the global average near-surface atmospheric temperature increased by 0.74 °C, with the rate of climate change in the past 50 years

Y. Lu, *Urban Form and Urban Heat Island*, Urban Sustainability,
https://doi.org/10.1007/978-981-96-3541-2_1

being twice that of the previous 100 years. Projections indicate that by 2081–2100, global temperatures will climb even higher.

In the context of global warming, rapid urbanization has intensified the urban heat island effect. The "World Urbanization Prospects (Revised 2014)" report by the United Nations Department of Economic and Social Affairs shows that from 1950 to 2014, the global urban population grew from 746 million to 3.9 billion, accounting for 54% of the world's total population. By 2050, the urban population is expected to increase by another 2.5 billion, reaching 66% of the global population [1]. Research by the U.S. Environmental Protection Agency indicates that for every million increase in urban population, the average annual urban–rural temperature difference during the day rises by 3 °C, and by 12 °C at night [2].

1.1.2 Heat Island Effect Has Become a Significant Factor in Worsening Urban Climate Conditions

Accompanying rapid urbanization is the drastic transformation of city environments and the widening urban–rural temperature gap. In European cities, this temperature difference exceeds 10 °C compared to surrounding rural areas [3], and even small cities exhibit a heat island effect [4]. The intensive utility of air conditioners and increased water consumption indicate that urban areas endure higher temperatures than rural regions [5, 6]. In Shanghai and Tianjin, two main cities in China, the rapid expansion of urban construction land has directly contributed to the heat island effect. Scholars have attributed this enhancement of the heat island effect to changes in land use and land cover, which alter the heat storage capacity of urban areas. These changes include the densification of urban spaces, the artificialization of urban surfaces, and significant anthropogenic heat emissions released by human activities. These alterations have profoundly impacted the regional atmospheric environment, boundary layer structure, and climate conditions, exacerbating the heat island intensity and urban energy consumption, thereby creating a vicious cycle. Chen et al. [7] analyzed temperature data provided by meteorological stations from 1960 to 2005 in the Wuhan metropolitan area, demonstrating a strengthening heat island effect and intensified warming since the early twenty-first century [7].

For most cities worldwide, the urban heat island effect increases human discomfort, health problems, energy consumption, and air pollution. In the context of global climate change, the heat island effect is poised to become a significant factor in reducing urban livability. Studies indicate that substantial heat emissions heighten health risks for urban residents [8]. Moreover, rising urban temperatures elevate surface ozone levels, directly impacting human health [9, 10].

1.1.3 Urban Planning and Management Require Refined Strategies Aiming at Mitigating Urban Heat Island

Firstly, the refinement of urban planning management is reflected in the shift of regulatory focus from "incremental management" to "form management". The American planning academia has concluded over the past thirty years that "urban growth is uncontrollable, urban form is moldable." This concept is central to refined and scientific management in China's current development stage, which balances quantity and quality. Urban sprawl, representing quantity increase, reflects the socio-economic demand for urban land development. This demand is objective, and while the government can control or encourage increments, market demand itself does not shift due to government intervention. The government's role should be to facilitate the market's function. Therefore, the government should accommodate the socio-economic demand for the expansion of urban areas.

"Urban form" refers to the external configuration characteristics of the urban built-up area within a defined total amount. Different urban forms can arise from the same urban scale. Although the city government cannot determine the socio-economic demand for land development, it can plan and guide the final urban form to achieve harmony between people and land, environmental friendliness, and climate adaptability. As the government enhances its market service function, the "incremental management" approach, rooted in a planned economy ideology, will gradually be replaced by "form management." The government now focuses on regulating urban form under objective market demand to achieve planning goals.

Secondly, refined planning management requires higher technical regulation for urban form. Traditional urban form management is based on human environmental cognition, transforming perceptions of "beauty" into the basis for managing urban form. It also considers basic technical factors such as lighting, fire protection, and sightlines, proposing control indicators with specific dimensions and threshold ranges [11]. As people demand higher urban environmental quality, the energy efficiency, safety, and climate adaptability of urban spaces have gained attention, leading to more detailed quantitative requirements for urban form governance.

1.2 Research Scope and Concept Definition

1.2.1 Research Scope

In recent years, the rapid urbanization of Wuhan has led to significant urban expansion and construction, exacerbating the urban thermal environment and the spread of urban heat islands (Xie et al. 2016) [12]. This book focuses on the 589 regulatory planningunits within Wuhan's main urban area, encompassing a total area of 522.68 km^2 (Fig. 1.1). This area aligns closely with the main urban area of Wuhan and includes most of the built-up areas, offering a diverse range of urban form types (Figs. 1.2 and 1.3).

All regulatory planning management units within this research scope serve as the study's samples. The plan is to use these units as statistical objects, calculating their average heat island intensity and various urban form indicators to explore the correlation between heat island intensity and urban form. The regulatory planning units were chosen because most regulatory planning units are between 0.3 and 0.5 km^2, which is the smallest spatial unit for urban planning and management, and represent the basic scale of urban spatial mechanisms. Studying at the regulatory planning management level facilitates the connection between urban planning strategies and land use regulation. Also, previous research has shown that using regulatory planning units for urban microclimate studies is feasible [13].

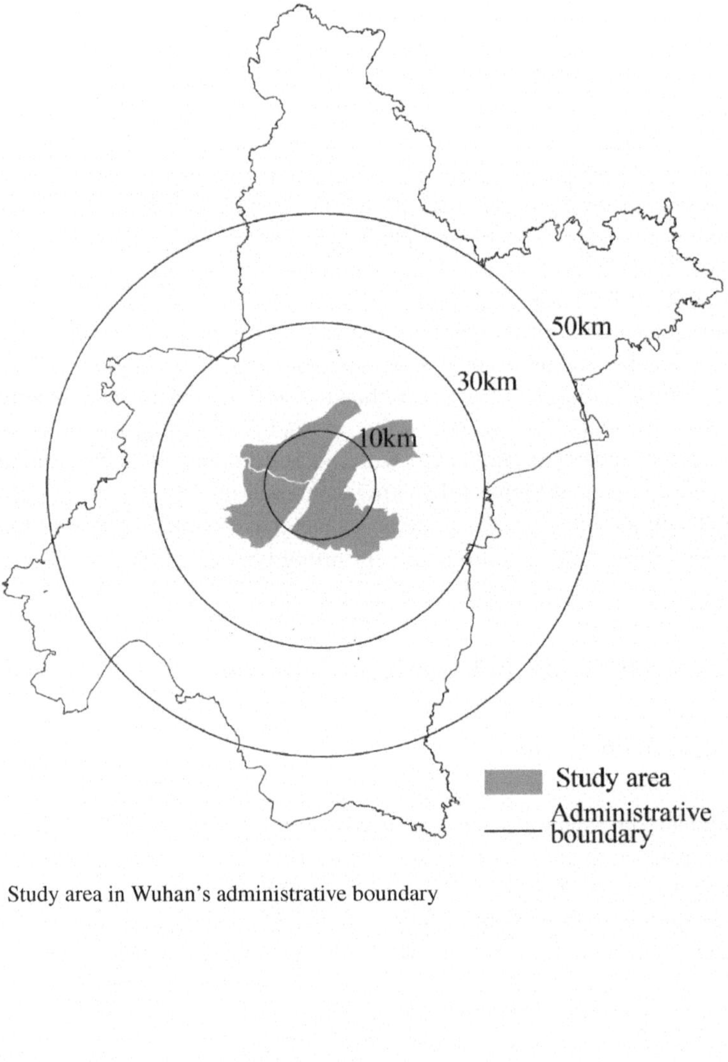

Fig. 1.1 Study area in Wuhan's administrative boundary

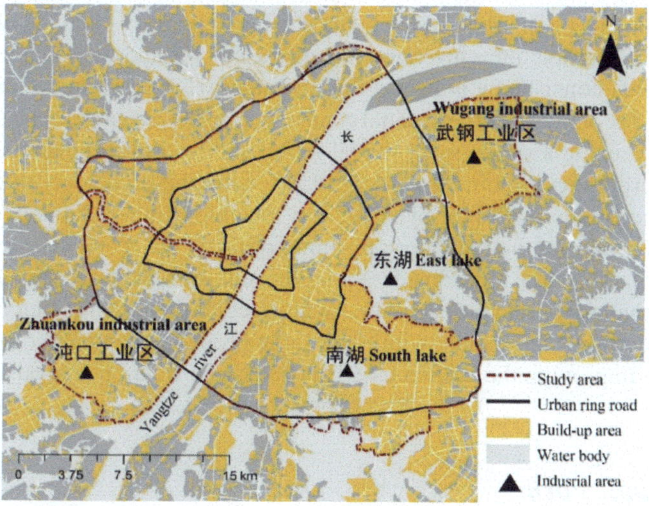

Fig. 1.2 Built-up area within study area

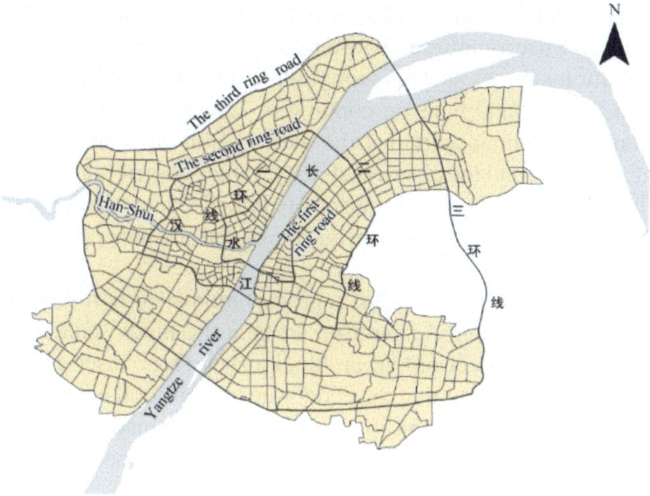

Fig. 1.3 Regulatory planning units in study area

However, defining sample boundaries involves the modifiable areal unit problem (MAUP), which means analysis results can change in line with the basic area unit definition [14]. This book argues that using regulatory planning units as spatial samples would offer rationality in both scale effects and zoning effects, which therefore has advantages over grid sampling (Fig. 1.5). The detailed reasons for applying regulatory planning units as spatial grids are as follows. Firstly, the spatial pattern of urban heat islands in Wuhan's main urban area, obtained using regulatory planning units, aligns with the surface temperature inversion pattern based on Landsat 8 data with a 30 m resolution. This consistency indicates that the analysis results are not distorted by scale effects. Secondly, the urban spatial context within the research scope is complex. Consequently, using a single-scale spatial grid could lead to significant differences in spatial morphology within the same grid, affecting the objectivity of correlation research conclusions. Dividing by regulatory planning units helps alleviate these issues because, during the regulatory planning process, the road network and property boundaries are taken into consideration. This minimizes the division of main and secondary urban roads and property areas within regulatory planning units, ensuring spatial morphology consistency within the same property range and road-enclosed plot, thereby improving the accuracy of This book's correlation analysis.

1.2.2 Urban Form

In a seminal work "A Theory of Good City Form," posits that urban form encompasses the physical environment, including the spatial organization of permanent objects such as buildings, streets, facilities, hills, and rivers. Considering these objects involve various attributes like use, quality, and ownership, the urban form requires integrating multiple information sources such as topographic data, land use information, street signs, facility networks, and housing conditions. Lynch defines "urban form" as a spatial arrangement of human activities, the spatial result of the movement of people, goods, and information, and the physical characteristics describing spaces significant to these activities. Thus, urban form is intricately linked to human activities, including both the man-made physical environment and the social attributes arising from human activities.

This article focuses on the physical representation of urban blocks, exploring how the external form elements of built-up areas at the block scale affect the local heat island effect. The concept of "urban block urban form" used here is narrower than Lynch's "city form," excluding the economic and cultural attributes of physical space. It extends beyond mere "morphology" (the combination features of block buildings) to encompass the appearance features and land use functions of blocks comprehensively.

Additionally, it is crucial to note that the term "urban block urban form" in this book emphasizes the enclosed areas defined by main and secondary urban roads. The delineation of regulation planning regulatory planning units considers the integrity of urban spatial texture, with clearly defined boundaries between different units, providing a solid statistical basis for this research.

It is particularly worth noting that for the convenience of writing, this study refers to "urban block urban form" simply as "urban form".

1.2.3 Heat Island Effect

The American Meteorological Society defines an urban heat island as a relatively warm area with closed isotherms, commonly found in urban areas. The heat island effect refers to the intensity of this phenomenon, measured by the temperature difference between urban and suburban areas. This effect varies significantly with time and location, with measurement indicators divided into surface temperature and near-surface atmospheric temperature, ranging in scope from the city level to the block level.

In this book, the "heat island effect" specifically refers to the Surface Urban Heat Island (SUHI) effect (Weng 2009) [15, 16]. This is defined as the difference between urban surface temperature and the average surface temperature of suburban farmland (see Sect. 3.1.2 for specific calculation methods). A positive value indicates a heat island effect, where the urban area is warmer than the suburbs, while a negative value indicates a cold island effect.

The focus on the surface heat island effect, rather than the atmospheric temperature heat island effect, is due to the need to measure heat island intensity across different areas simultaneously to ensure data comparability. The limited number of temperature measurement points in Wuhan, most of which are far from built-up areas and probably merely influenced by urban form, makes it difficult to obtain a sufficient number of valid samples. Land surface temperature data obtained via remote sensing covers a wide range and ensures consistent measurement times across different areas.

Furthermore, existing studies show a significant correlation between land surface temperature and air temperature under light wind conditions [17–20]. Thus, as long as wind speeds are low on the measurement day, both types of data are valid for studying the correlation between urban form and heat island effect. Therefore, using land surface temperature is more conducive to the objectives of this book.

1.3 Data Source

1.3.1 Land Surface Heat Island Intensity

The Landsat 8 series of remote sensing images, provided by the United States Geological Survey (USGS), have undergone radiometric and geometric correction to achieve L1T level product quality. The projection for the Landsat 8 image of the Wuhan area is WGS 1984 UTM zone 49N.

The specific Landsat 8 image used in this book was captured at 10:55 Beijing time on October 6, 2014 (Table 1.1). The selection of this image was based on several considerations:

Table 1.1 Meteorological observation data of Wuhan from October 1, 2014, to October 6, 2014

Date	Weather	Temperature	Wind direction
October 1, 2014	Cloudy/cloudy	26 °C/18 °C	No sustained wind direction $\leq$ level 3
October 2, 2014	Sunny/sunny	28 °C /16 °C	No sustained wind direction $\leq$ level 3
October 3, 2014	Sunny/sunny	29 °C/17 °C	No sustained wind direction $\leq$ level 3
October 4, 2014	Sunny/cloudy	29 °C/16 °C	No sustained wind direction $\leq$ level 3
October 5, 2014	Cloudy/sunny	28 °C/16 °C	No sustained wind direction $\leq$ level 3
October 6, 2014	Sunny/sunny	27 °C/12 °C	No sustained wind direction $\leq$ level 3

1. **High Spatial Resolution**: Landsat 8 provides high-resolution surface temperature data, the highest among currently available open-source remote sensing data, meeting the research requirements of this book.
2. **Minimal Cloud Cover**: After screening various images, this particular one had minimal cloud cover over the research area, making it the highest quality image available for that year.
3. **Significant Daytime Temperature Variation**: Surface temperature exhibits greater spatial variation during the day, while air temperature varies more at night. The selected daytime surface temperature data is conducive to distinguishing differences in heat island intensity across various areas, providing an ideal study area regarding the relationship between spatial morphological features and heat island intensity [21].
4. **Light Wind Conditions**: Existing research shows a significant correlation between surface and air temperatures under light wind conditions. On the day the satellite data was collected, Wuhan experienced wind speeds below level 3. Thus, even though land surface and air temperatures are not equal, their relative relationship and spatial variation remain consistent.
5. **Timing and Focus**: Although the data collection time is not at the peak temperature of the day (usually around 13:00) nor the peak summer temperature, the study focuses on the impact of spatial morphology on land surface temperature variation rather than absolute temperature values. Therefore, the timing does not affect the analysis results.
6. **Weather Conditions**: The weather in Wuhan was clear for the five days preceding October 6, 2014, with no rainfall and light wind, ensuring a stable thermal environment on the data collection day.

Since the land surface temperature data used in this book is collected at a specific point in time, it must be compared with land surface temperature inversion results from other times to demonstrate its typicality. To achieve this, this book utilizes MODIS satellite remote sensing data for comparison. Unlike the high-resolution Landsat 8 data, MODIS offers the advantage of capturing surface temperatures at multiple time points, although its resolution is lower.

The MODIS data used in this book includes the "MODLT1M China 1KM Surface Temperature Monthly Synthetic Product," which is obtained from https://www.gscloud.cn/sources/details/336. The land surface temperature monthly synthetic products for each month from June 2014 to October 2014 were extracted (Fig. 1.4). It was found that the high-resolution surface temperature distribution captured by Landsat 8 on October 6, 2014, aligns well with the summer monthly temperatures of 2014. This indicates that the Landsat 8 data has good typicality and can represent the spatial variation of land surface temperature in Wuhan during the daytime on extremely hot days.

1.3.2 Urban Form Indicators

The data sources for urban form indicators include:

1. Land Use dataset: This map of Wuhan's main urban area is used to calculate the proportion of various land uses within each management unit.
2. Three-dimensional Building dataset: Also from Wuhan's main urban area, this map helps determine building density, floor area ratio, GSVF (Green Space View Factor), and other relevant indicators for each management unit.
3. Land cover dataset: This data is retrieved from the same source used for calculating land surface temperature. Through supervised classification, we retrieve the proportion of different types of surface cover in each management unit from Landsat 8 Satellite Remote Sensing Data
4. National Census Data: The sixth national census data is used to calculate the permanent population density for each management unit.

1.4 Research Purpose and Significance

1.4.1 Research Purpose

1. **Reveal the mechanism of urban form on the heat island effect**

Start with a theoretical analysis to build a comprehensive framework that explains how various urban form indicators influence the heat island effect. Combine the theoretical framework with quantitative analysis to systematically explore and verify the impact of different urban form indicators on the heat island effect. This includes examining the spatial effects and providing a detailed explanation of the mechanisms involved.

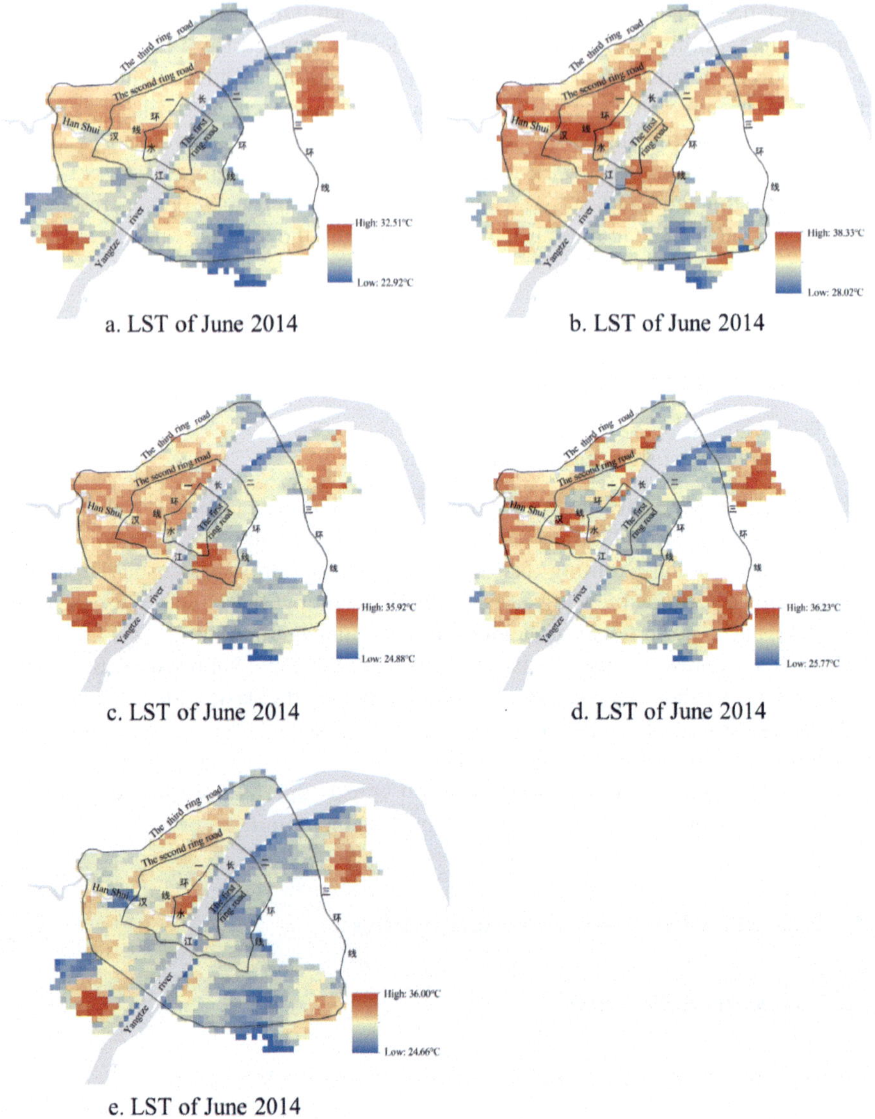

Fig. 1.4 MODIS based monthly mean surface temperature distribution for some months of 2014 and Landsat8 surface temperature distribution for October 6, 2014. *Source of information* Geospatial Data Cloud

2. **Propose urban form optimization techniques to reduce heat island intensity**

Based on the identified mechanisms through which urban form indicators affect heat island intensity, we develop techniques and methods for controlling these indicators. We use the findings to guide urban planning practices, ensuring that urban form optimization techniques are effectively implemented to reduce the intensity of urban heat islands.

By achieving these objectives, this book aims to provide a solid foundation for understanding the relationship between urban form and the heat island effect and to offer practical solutions for mitigating the impact of urban heat islands through informed planning and design strategies.

1.4.2 Research Significance

1. **Theoretical significance**

Traditional spatial analysis has experienced three primary periods. Initially, researchers focused on the physical environment, which emphasized aesthetic principles and formal order. This perspective sought to understand urban spaces based on visual appeal and structural harmony. Subsequently, researchers shifted from to physical environment to the cognitive environment. the relevant studies explored how individuals perceive and interact with their surroundings, focusing on environmental perception and human behavior. This approach extended beyond mere aesthetics to consider how people mentally experience and sense urban spaces. The analysis evolved to encompass the regional environment, emphasizing spatial regionalism and the complex relationships between humans and land. This dimension addressed material environments, human behavior, and the socio-economic forces shaping urban spatial organization [22].

While cognitive environment theories, such as Kevin Lynch's "City Image" and Norberg-Schulz's "Existential Space" theory, advanced the understanding of urban from an individual psychological perspective, they predominantly explored the "subjective impact" of spatial environments. These theories focus on how people perceive and experience urban spaces. However, they often overlooked the "objective impact" of urban forms based on scientific and empirical evidence. In detail, in the context of global warming and the increasing prominence of the urban heat island effect, traditional theories have not sufficiently addressed the relationships between urban form elements and local microclimate performance indicators.

To bridge this gap, this book introduces a novel perspective by analyzing urban form through the lens of engineering science. This approach shifts the focus from subjective experiences to the objective impact of urban forms on local microclimates. Specifically, This book investigates the physical properties of urban forms and their effects on the local heat island effect, offering a new theoretical framework for urban form theory. By examining how specific urban form elements influence microclimate conditions, this research aims to enrich the existing body of knowledge with empirical evidence and scientific analysis.

In summary, this book represents an innovative approach by addressing the objective relationships between urban form and microclimate effects, which is crucial for developing effective urban planning strategies in the era of global climate change.

2. **Significance in urban planning practical**

The ability to effectively respond to climate change is a crucial factor for the sustainable development of future cities. This book aims to enhance our understanding of the causes behind the urban heat island effect and to provide a scientific foundation for promoting healthy and sustainable urban development.

In recent years, urban planning practices in major cities have increasingly focused on strategies to mitigate the urban heat island effect. This effect, where urban areas become significantly warmer than their rural surroundings, is a major concern for urban sustainability. As cities continue to expand, planners and researchers are seeking effective methods to manage this phenomenon and improve urban environments. For instance, the 2006 Shenzhen Urban Master Plan revision introduced principles for enhancing urban natural ventilation. This plan emphasized the creation of ventilation corridors and optimal spatial layouts for various urban structures to improve air flow and reduce heat island effects. Similarly, in 2010, Changsha City developed the "Changsha City Urban Ventilation Planning Technical Guidelines," which outlined technical measures for urban ventilation, addressing commercial, residential, and industrial areas.

In 2014, Beijing undertook a comprehensive study to design a network of urban building wind tunnels, establishing a system of ventilation corridors. The following year, Wuhan developed a general layout plan aimed at mitigating the urban heat island effect through a combination of ecological green spaces and strategic ventilation. This plan proposed eight "primary compensation areas," five "main wind tunnels," and twelve "key action areas" to address the issue of heat island effect mitigation and improve the city's environmental conditions.

These planning practices reflect a broader trend of integrating ecological spaces and ventilation strategies to manage the urban thermal environment. However, there is also a pressing need to focus on the built environment itself. This involves regulating the physical elements of urban spaces to reduce heat island intensity, a challenge with significant implications for urban planning.

This book provides practical guidance for optimizing urban forms to reduce heat island effects. It offers valuable insights for:

1. Optimizing Urban form: The findings of this book are applicable to the design of overall urban layouts that balance environmental quality with functional urban spaces.
2. Formulating Control Regulations: The research supports the development of more effective planning regulations aiming at mitigating heat island effects through strategic urban design.

3. Refining Planning Practices: The study offers a foundation for fine-tuning planning approaches to address the challenges of urban heat islands and promote sustainable urban development.

By focusing on both theoretical analysis and practical solutions, this research contributes to the advancement of urban planning practices that can adapt to and mitigate the impacts of climate change.

1.4.3 Key Scientific Issues to Be Addressed

1. **Mechanisms of urban block form affect the urban heat island effect**

Correlation and causality are two closely related but distinct logical propositions. Some studies have conflated the two concepts when exploring the relationship between urban form and the urban heat island (UHI) effect, using correlation to substitute for causality, leading to logical fallacies. In this book, the UHI effect, as an outcome, is influenced by multiple factors and its formation mechanism is complex. Moreover, due to the lack of experimental conditions, it is impossible to eliminate confounding variables. Therefore, this book seeks to explain the influencing mechanisms not only based on the findings of quantitative correlation studies but also by incorporating principles of heat transfer to propose how urban form indicators affect the UHI effect from a physical perspective. Furthermore, since urban form indicators vary widely in type and attributes, they have both positive and negative impacts on the UHI effect. Therefore, a major challenge of this book is determining how to combine the effects of different urban form indicators and ultimately assess the comprehensive impact of urban form on the UHI effect.

2. **Urban form control techniques for reducing urban heat island intensity**

Mechanism research can reveal how different urban form indicators influence UHI intensity, but the development of urban form control techniques requires more quantifiable control targets and more targeted, differentiated approaches. Based on the exploration of the relationship between urban form and the UHI effect, a second key issue of this book is how to translate the conclusions of correlation studies into practical urban form control techniques aimed at reducing UHI intensity.

1.4.4 Research Content

1. **Identification of the spatial patterns of the urban heat island effect**

This section addresses the question, "What are the spatial distribution characteristics of the urban heat island (UHI) effect?" First, high-resolution land surface temperature (LST) data (30 m × 30 m) will be retrieved using Landsat 8 remote sensing data to identify the overall spatial pattern of the UHI effect. Then, based on the LST variations, different UHI intensity zones will be distinguished, and the spatial distribution patterns of each intensity zone will be analyzed.

2. **Core urban form indicators influencing the UHI effect**

This section responds to the question, "What are the core urban form indicators affecting the UHI effect?" Given the vast number of urban form indicators, it is essential to identify the core indicators that significantly influence the UHI effect. First, potential urban form indicators will be identified based on relevant literature. Then, the spatial characteristics of typical high-intensity UHI zones and low-intensity UHI zones will be extracted and compared to discern the differences in urban form between different UHI intensity zones, leading to the selection of core urban form indicators affecting UHI intensity.

3. **Mechanisms of how urban block urban form indicators affect the UHI effect**

This section answers the question, "How do urban form indicators influence UHI intensity?" covering both the mechanisms and effects. First, based on existing literature and basic heat transfer theories, a logical framework will be established to explain the relationship between urban form indicators and UHI intensity. Then, a quantitative analysis will be conducted to explore the correlation patterns between various urban form indicators and UHI intensity, investigating both the intensifying and mitigating effects of these indicators as well as the modifying role of external environmental factors. Finally, the contribution of each urban form indicator to UHI intensity will be compared.

4. **Urban form control techniques for reducing UHI intensity in urban blocks**

This section addresses the question, "How can urban form indicators be regulated to reduce UHI intensity?" First, the thresholds at which various urban form indicators impact UHI intensity will be analyzed. Based on these thresholds, early warning and control techniques will be proposed, along with strategies for optimizing urban forms. Next, zoning will be conducted based on the previously discussed mechanisms, and differentiated control techniques will be proposed for each zone, considering the relationship between urban form indicators and UHI intensity in different areas. Lastly, specific urban form control techniques will be developed for particular urban functional zones.

1.4.5 Research Methods and Framework

1.4.5.1 Research Methods

Case Study Method

The case study method involves analyzing a single object to derive general and universal laws. In this book, typical cases are selected based on land surface temperature retrievals, including three typical strong heat island areas and three typical weak heat island areas. A detailed analysis of the urban form indicators for each case is conducted to summarize the urban form characteristics of both strong and weak heat island areas. Another instance of the case study method is applied to the control techniques of block-level urban forms. Through an analysis of the current building layout and wind-heat environment of the Hualou Street block, urban form indicator regulations are transformed into concrete optimized planning schemes.

Statistical Analysis Method

The statistical analysis method refers to using the quantitative results of various attributes of the study object to understand and reveal the relationships, patterns of change, and developmental trends among things, in order to achieve accurate explanation and prediction. In this book, the heat island intensity and urban form are statistically analyzed. After statistical analysis of each variable, correlation analysis and regression analysis are used to quantify the relationships between them.

Logical Induction Method

The logical induction method is based on a series of empirical judgments and data reserves to identify general patterns that things typically follow and hypothesize that other similar things also follow these patterns. This research method is used multiple times in this book. For instance, the core urban form indicators influencing heat island intensity are selected based on case analysis and a comparison of urban forms in areas with different heat island intensities. Additionally, the complex influence of urban form on heat island intensity is generalized into two effects: intensifying and mitigating, which are treated as universal rules.

1.4.5.2 Research Framework

See Fig. 1.5.

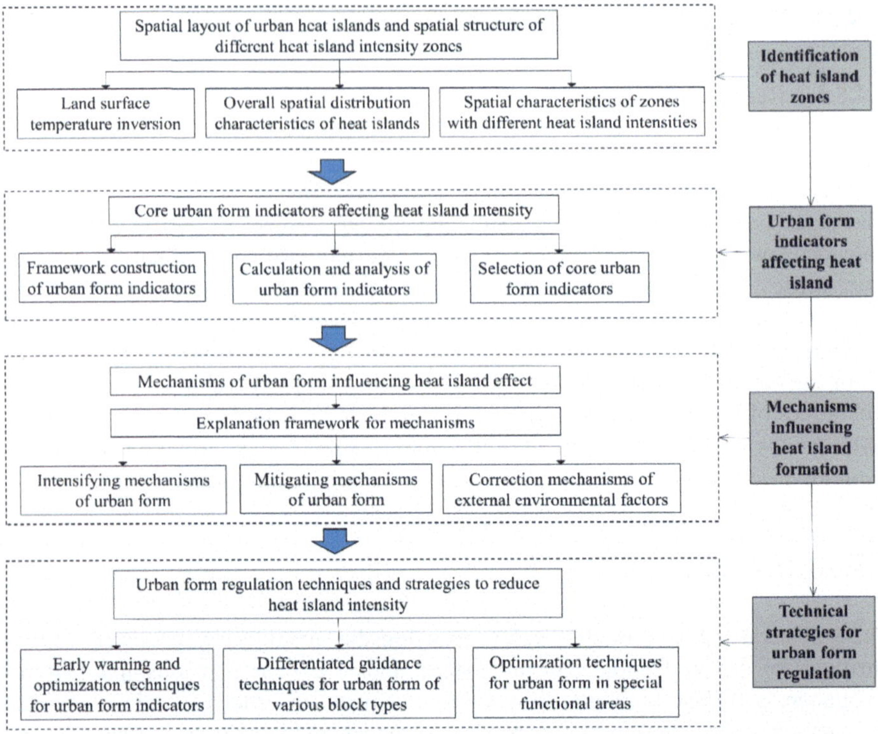

Fig. 1.5 Research framework

1.5 Literature Review

1.5.1 Basic Route of Urban Form Research

1.5.1.1 Physical Environment Research Based on Aesthetic Rules and Morphology Order

Physical environment research forms a crucial part of classical urban design theory. It extends architectural concepts to the urban space level and serves as the foundation for analyzing urban material spaces. This research approach examines the physical components of urban spaces and their structural order, addressing both the elements that make up urban environments and their morphology organization (Table 1.2).

Urban space primarily consists of two elements: squares and streets. In contrast, Hamid Shirvani expanded on this idea by detailing a more comprehensive set of urban space elements. His framework includes land use, building form and volume, traffic and parking, open spaces, pedestrian pathways, and supporting activities. Additionally, the Japanese publication "New Architecture Series 17. Urban Design"

Table 1.2 The main context of the existing literature regarding to urban form

	Representative characters	Theoretical principle	Philosophy
Research on entity environment	R· Creaer Hamid Sherwar Seater A. Jacobs	Classical urban design theory	Urban space is composed of different physical elements; the order of urban form follows the aesthetic law and physical order
Cognitive environment research	Kevin Lynch Norberg Schultz Aldo Rossi	Environmental metaphysics	Material space environment is the result of human subjective experience, and the urban form needs to be understood from people cognitive law
Research on regional environment	Jones Conzen Smailes	Urban geography	Urban spatial form is composed of different landscape areas; spatial growth follows the laws of "hyperplasia" and "substitution"

shifts the focus from the mere "quantity" of urban form to its "quality." It emphasizes that the intensity and function of urban spaces are integral aspects of urban form.

Urban composition theory is a significant branch of urban form theory that explores the relationships between urban elements to create aesthetically pleasing environments. This theory delves into the beauty of spatial structures, examining principles such as symmetry, balance, and harmony among urban components [22]. Camillo Sitte's work aligned with this theory, focusing on the functional associations between different urban elements and their overall coherence within the city's material space.

In 1987, Jacobs and Appleyard proposed a framework for a livable urban environment that encompasses five core features: well-designed streets and neighborhoods, minimal residential density and land use intensity, diverse and appropriately located activities, well-defined public spaces, and a unique architectural mix. This framework highlights the importance of interactions among urban space elements and emphasizes creating a well-structured and aesthetically cohesive urban environment.

A common theme in these theories is the emphasis on the material aspects of urban space while often overlooking the human experiences and psychological impacts of these spaces. The focus on "seeing things but not people" reflects a tendency to analyze urban environments through a purely physical lens, without fully considering the ways in which human behavior and perception interact with these environments.

1.5.1.2 Cognitive Environment Research Based on Environmental Cognition and Human Behavior

Another important dimension of urban material space analysis focuses on human cognition, considering the material environment as a result of subjective human experiences.

In 1960, Kevin Lynch introduced a groundbreaking approach to analyzing urban forms by exploring how people perceive and remember their environments. Lynch's work centered on the idea of urban legibility, which refers to the ability of a city's physical features to create strong and memorable mental images for its inhabitants. By asking residents to draw simple maps of their cities, Lynch identified five core elements of the cognitive environment: roads, edges, landmarks, regions, and nodes. These elements form the basic components of how people mentally organize and navigate urban spaces.

Building on Lynch's work, Norberg-Schulz sought to establish a more systematic method for understanding urban spaces through his concept of "existential space." Schulz proposed that urban spaces are structured hierarchically, with each level of the hierarchy characterized by distinct spatial features. He described this structure as having two main components: the vertical structure, which spans from large geographical regions to individual homes, and the horizontal structure, which includes domains (areas of special significance), paths (routes connecting different centers), and places (centers of importance). This framework aimed to create a more stable and recognizable system for understanding urban spaces.

Aldo Rossi (1966) further advanced the discussion by exploring the historical and cultural dimensions of urban spaces. Rossi argued that traditional architectural forms and spaces serve as symbols of collective memory and cultural heritage. He believed that the city functions as a repository of historical and cultural narratives, reflecting the "prototype" of regional architectural traditions and the cultural identity of its inhabitants. Rossi's approach emphasized that urban spaces are not merely physical environments but also carriers of historical and cultural significance.

While these theories moved beyond the traditional view of urban space as simply a physical entity, they generally offered broad and sometimes vague definitions of urban elements. These approaches aimed to incorporate subjective experiences and cultural meanings but often lacked the practical specificity needed for direct application in urban planning and design.

In summary, these perspectives introduced new ways of understanding urban spaces by highlighting the role of human perception and cultural context. However, they remain somewhat general and less directly applicable to practical urban planning challenges.

1.5.1.3 Regional Environment Research Based on Spatial Regionalism and Human-Land Relationship

Research on urban space from a geographical perspective examines the two-dimensional physical layout of cities, focusing on the internal spatial differentiation of urban environments.

Jones's approach to urban space divides the physical environment into five distinct types based on factors such as construction year, usage function, and architectural form. By analyzing these categories, Jones identified various spatial distribution patterns of urban landscapes, highlighting how different building types and functions shape the city's physical form over time.

Building on this foundation, Louis explored the basic characteristics of three different levels of urban edge zones: the old city, early suburbs, and main residential areas. His analysis revealed the geographical evolution of urban spatial morphology, showing how cities expand and develop across different historical periods.

In a similar vein, Conzen [23] introduced the concepts of "edge zones" and "consolidation boundaries". Conzen proposed that the evolution of urban physical space progresses through three distinct phases: an acceleration period, a deceleration period, and a stable period. During these phases, urban development is initially driven by rapid growth, which eventually slows down and stabilizes as it encounters physical or regulatory boundaries. Over time, however, these boundaries are often overcome, leading to new periods of expansion. Smailes [24] further developed the understanding of urban spatial dynamics by proposing a dual process of urban form evolution. He described urban growth as a combination of "proliferation", or outward expansion into new areas, and "replacement", or internal reorganization of existing spaces. This dual process illustrates how cities continuously evolve both spatially and functionally.

These studies collectively enhance our understanding of how urban spaces evolve geographically. They provide insights into the patterns of urban growth and transformation within cities, emphasizing both the physical and organizational aspects of urban development.

1.5.2 Measurement of Urban Heat Island Effect

1.5.2.1 Measurement Indicators

The definition of the heat island effect in this book centers on measuring spatial variations in temperature. Existing literature identifies several key indicators for quantifying urban heat island intensity, each providing different insights into the phenomenon. These indicators can be broadly categorized into three types: measuring absolute temperature differences, assessing the size of high-temperature areas, and evaluating variations compared to background temperatures.

Indicators for Measuring Absolute Temperature Differences

The most fundamental method for measuring the urban heat island effect is comparing absolute temperature values. This approach typically involves selecting a reference area and comparing temperature measurements at the same time and height across different locations. Oke and East [25] demonstrated that comparing temperatures in urban and suburban areas is an effective way to study the heat island effect.

Traditionally, suburban farmland areas serve as the reference points for these comparisons [26, 27]. By analyzing multi-year meteorological data from urban and suburban stations, researchers can assess the relative intensity of the heat island effect and observe both seasonal and daily variations [28]. Some studies have refined this method by examining specific temperature differences, such as between city centers and rural areas [29] or between urban built-up areas and water bodies [30]. Additionally, methods that average temperatures from multiple stations, like Liu [31] approach using four urban and three suburban stations, offer a more objective measure of heat island intensity. Similarly, eight temperature stations were utilized in Shenzhen to gauge the heat island effect through average temperatures.

Indicators for Measuring the Size of Hot Areas

With advancements in satellite remote sensing technology, researchers can now capture high-resolution spatial distributions of urban surface temperatures over larger areas. This technology enables more detailed observation of the heat island effect. Traditionally, equal temperature divisions have been used to map high-temperature areas, but some researchers argue that using standard deviations for defining heat island zones provides more accurate results.

For instance, Streutker [32] employed two-dimensional Gaussian distribution functions to extract urban heat island areas from surface temperature data. Zhang and Wang [33] proposed a step-by-step method to determine heat island areas: first, calculate the average surface temperature and its standard deviation for the city, second, use the following formula to calculate the threshold of the hot area.

$$\mathrm{T} = \mathrm{a} + \mathrm{x} \times \mathrm{sd}$$

where a is the average temperature, sd is the standard deviation, and x ranges from − 2.5 to 3) to set temperature thresholds; third, categorize surface temperatures into 11 levels; and fourth, analyze the number of pixels and their distribution for each temperature category (Fig. 2.1). This method offers a standardized approach for comparing heat island areas across different urban environments.

Indicators for Measuring Variation Compared to Background Temperature

Another approach to studying the heat island effect focuses on the difference between local temperatures and background temperatures. While traditional definitions of rural areas as unaffected by the urban heat island effect are somewhat vague [20], this approach seeks to quantify how specific urban temperatures deviate from a broader background temperature.

Imhoff et al. [29] and Zhang et al. [34] analyzed urban heat islands by comparing the average surface temperature of urban areas to that of a 20 km^2 background area. To reflect small-scale temperature variations, some researchers measure the proportion of areas where the surface temperature exceeds a certain threshold [35]. Kuang et al. [36] developed the Urban Heat Field Variation Index, calculated using the formula

$$\mathrm{HI} = \frac{\mathrm{T} - T_{mean}}{T_{mean}}$$

where HI represents the heat island variation index, T is the observed surface temperature, and Tmean is the average temperature of the research area.

Choosing appropriate heat island intensity measurement indicators is crucial for obtaining accurate research results. Schwarz and Manceur [37] explored different methods for calculating heat island effects and found that varying measurement techniques affect the correlation between urban morphology and heat island intensity. He identified three primary methods: ① comparing average temperatures between built-up and non-built-up areas, ② measuring the difference between peak and average temperatures of built-up land in urban areas, and ③ using Gaussian distribution to map heat island areas. Schwarz concluded that different methods capture different attributes of the heat island effect and that selecting the right indicators is essential for understanding urban heat dynamics (Table 1.3).

1.5.2.2 Measurement Methods

Since Howard's pioneering work over a century ago, the study of the heat island effect has evolved significantly. Based on a review of scholarly methods for observing the UHI effect, this article categorizes and summarizes the main approaches used in contemporary research.

Meteorological Data Method

The meteorological data method involves analyzing UHI dynamics through fixed-point observations of meteorological data from urban and suburban locations. This method has been foundational in UHI research and continues to be widely used. Kłysik and Fortuniak [40] and Morris et al. [41] studied UHI effects in cities like Melbourne, Fairbanks, and Łódź, demonstrating that UHI intensity is influenced by factors such as cloud cover and wind speed, with the strongest effects observed on clear, windless nights. In China, suburban meteorological data was used to analyze the heat island effect in Shanghai. This method offers long-term data for analyzing temporal changes in UHI intensity but is limited by the number and placement of meteorological stations, which can affect data accuracy and spatial coverage.

Table 1.3 Urban heat island intensity indicators

Metric	Unit	Quantitative method	Pertinent literature
City-agricultural district	K	The difference between the average surface temperature in urban areas and farmland	
Core area-country	K	The difference in mean surface temperature between the urban administrative region and the surrounding 20 km^2	[29] [34]
City-other areas	K	The temperature difference between the city and all the other areas	[38]
City-water area	K	The difference in the surface temperature between cities and water surfaces	[30]
Gaussian area	Km^2	After removing the suburbs, the area where the surface temperature is distributed according to a two-dimensional Gaussian function	[32]
Gaussian intensity	K	The highest temperature in the Gauss region	[32]
Heat island area	Km^2	The surface temperature is above a region that is one standard deviation above average temperature	[33]
Intensity	K	The difference between the area of the highest value and the average value of the surface temperature after excluding the water area	[39]
Micro heat island	%	The portion of areas where non-water surface temperatures are higher than the highest trees	[35]
Standard deviation	K	Standard deviation of the surface temperature in the administrative region	
The empirical value for the gaussian intensity	K	The difference between the maximum rural surface temperature under the urban and Gaussian distribution curves	

Mobile Observation Method

The **mobile observation method** uses vehicles equipped with meteorological instruments to study local microclimates. This approach addresses some limitations of fixed observation stations and provides flexible, dynamic data collection. Mobile observations were used to study the heat island effect and the cooling effects of green spaces in Singapore. Guo et al. [42] employed mobile methods with vehicle-mounted instruments and GPS to investigate the impact of different urban surfaces on the local microclimate. This method is advantageous for capturing spatial variability and real-time data, but suffers from limitations in route comparability and data consistency due to varying observation conditions.

Remote Sensing Inversion Method

The **remote sensing inversion method** leverages satellite technology to measure land surface temperatures (LST) over large spatial scales, providing a broad perspective on UHI effects. Since the first use of the thermal band of the TIROS-1 satellite in 1972, remote sensing technology has advanced significantly. Researchers have used various satellite sensors (e.g., NOAA AVHRR, MODIS, Landsat TM/ETM+, ASTER) to study urban heat islands. This method involves extracting LST from thermal infrared images, analyzing radiative properties of urban surfaces, and interpreting the data to map heat distribution [26]. Remote sensing data facilitates the analysis of UHI effects on a macro scale but may be limited by factors like sensor resolution and data interpretation challenges.

Model Simulation Method

The **model simulation method** employs mathematical models to simulate UHI dynamics and analyze heat island effects based on observational data and theoretical frameworks. Various models were used, including climate statistical models, urban boundary layer models, and energy balance models. Also, researchers studied the effects of urban terrain on temperature distribution. These models are used for analyzing the energy balance of urban surfaces, exploring heat exchange processes, and predicting UHI changes under different scenarios (Yang et al. 2003). While highly effective for detailed analysis, these methods require substantial computational resources and complex data inputs.

Comparison of Air Temperature Heat Island and Surface Temperature Heat Island

Air temperature heat islands and surface temperature heat islands represent two distinct aspects of UHI effects. Research has shown both correlations and differences between these two types of heat islands. Studies such as those by Arnfield [26] and Nichol et al. [43] have found that while air temperature and surface temperature heat islands often correlate, they do not always follow the same patterns over time. Air temperature heat islands are generally more pronounced at night, whereas surface temperature heat islands are usually strongest during the day [44]. It is crucial to differentiate between these two types of heat islands to fully understand UHI effects and develop appropriate mitigation strategies [45].

1.5.3 *Research on the Formation Mechanism of Urban Heat Island Effect*

In recent decades, scholars have conducted profound and extensive research on the temperature differences between cities and rural areas. Since Luke Howard (1833) published academic achievements on London's urban climate from 1818 to 1833, the urban heat island effect is one of the most concerning phenomena in urban climatology. The formation process of urban heat islands is relatively complex, involving multiple disciplines, and can mainly be attributed to atmospheric circulation, local climate factors, and human factors, and is also related to factors such as urban population size and built-up area [29].

1.5.3.1 Explanation Based on Urban–Rural Differences

The comparison between urban and rural areas has led scholars to develop theories explaining the UHI effect through heat transfer mechanisms, which include conduction, convection, and radiation. Scholars believe that heat convection and radiation are the primary modes of heat transfer affecting urban heat islands.

Urban environments raise temperatures in several ways. First, impervious pavements and buildings absorb solar energy and have higher thermal conductivity than soil and vegetation, leading to heat conduction as the temperature difference between the building surfaces and surrounding air increases. This warmer air then mixes with cooler air, causing convective heat transfer, and continues to raise air temperatures. Second, greenhouse gases like water vapor and carbon dioxide absorb and re-emit radiation energy, contributing to the heat island effect. This energy comes from both direct solar radiation and the reflection of solar rays off the earth's surface and buildings.

The unique spatial structure of cities also prevents the cooling effects of surrounding ecological areas from mitigating built-up area temperatures. Dense buildings alter aerodynamic parameters, increasing aerodynamic roughness and zero-plane displacement, which hinders the exchange of high-temperature urban air with cooler surrounding air. Additionally, materials like bricks and cement, which have higher specific heat capacities than soil, continue to release heat at night, while suburban soil cools rapidly.

Kuang et al. [36] summarized the causes of urban heat islands as follows: the unique properties of urban surfaces fundamentally alter the thermal conditions of the underlying surface; air pollution in cities absorbs solar energy and forms an insulating layer that returns energy to the city as atmospheric counter-radiation; cities emit significant amounts of heat through sources like power plants, boilers, and motor vehicles; the reduction of green spaces and water bodies decreases natural cooling; and stable external weather conditions reduce the natural convection of air.

Table 1.4 Influencing indicators on urban thermal environment

Level 1 indicators	Urban cushion surface	Atmospheric circulation	Solar radiation	Human activity intensity	Improving the ecological environment
Level 2 indicators	Building density	Local air pressure system	Sunshine	Density of population	Vegetation index
	Building floor area ratio	Air temperature	Direct solar radiation	Roading density	Vegetation cover
	Impermeable surface cover degree of ground temperature	Wind direction	Sun scattering radiation	Urbanization Level	Landscape diversity index
	Digital elevation (ground) model	Humidity	Ground radiation	Industrial zone density	
		Cloud amount		GDP	
		Visibility		Anthropogenic heat and pollutants	
		Precipitation			

Source Yu [46]

Yu [46] proposed an index system of urban thermal environment impact factors (Table 1.4), while other researchers categorized UHI influencing factors into meteorological conditions, thermal properties of urban surfaces, and human heat release. The planning of Tokyo's "Sea Forest" wind path identified twelve urban surface elements affecting the UHI effect, categorized into artificial heat production, surface coverage, building morphology, and building function.

Research shows that urbanization changes the physical properties of surfaces, transforming natural soil into impermeable surfaces like cement and asphalt, which increases surface temperatures. This transformation, along with differences in building height and density, alters surface reflectivity, specific heat capacity, and thermal conductivity, forming a heat field with higher temperatures than surrounding non-urban areas [47] (Table 1.5).

1.5.3.2 Explanation Based on Energy Conservation

With the development of modern science, scholars tend to explain the cause of the heat island effect from the dimension of energy conservation. As early as the nineteenth century, Luke Howard, the discoverer of the heat island effect, proposed a hypothesis about its cause. He believed that humans, animals, and fire contributed to the winter heat island effect. While the summer effect was attributed to the city's dry air and complex surface, which allowed it to absorb more solar energy. Gartland [48] suggests that Howard's theory remains largely correct. Modern science has identified

Table 1.5 Influencing factors on urban thermal environment

Type	Project
Human heat discharge	Building heat radiation
	Thermal radiation of the area cooling system
	Motor vehicle thermal radiation
	Railway thermal radiation
	Thermal radiation in commercial activities
	Human heat
	Human latent heat
Surface cover	Water heat radiation
	Bare soil, grassland thermal radiation
	Vegetation heat radiation
	Bitumen thermal radiation
	Thermal radiation in built-up area
Architectural form	Average building spacing
	Average building height
	Sky dome visibility
Building function	Floor area ratio of office buildings

two main factors for the heat island effect: first, man-made surfaces composed of dark materials absorb and store solar energy, with street canyons capturing reflected sunlight. Second, most building materials are impermeable, preventing rainwater from dissipating heat through evaporation and lacking the cooling effect of plant evaporation.

Oke and East [25] identified eight factors influencing the urban heat island effect, all of which impact the city's surface energy and radiation balance:

- The vertical surfaces of buildings cause heat radiation to be reflected multiple times within the building group, absorbing more sunlight.
- Limited visibility of the sky in street canyons restricts the reflection of heat radiation to the sky, retaining more infrared rays; the higher the height and density of the buildings, the more restricted the view of the sky.
- More heat is absorbed by buildings and paved surfaces and released with delay, not only due to the thermal characteristics of the material but also because sunlight and infrared rays are trapped, and air convection is slowed.
- Most absorbed solar radiation is converted into sensible heat rather than latent heat, due to paving and impermeable surfaces covering moist soil and vegetation.
- A large amount of sensible and latent heat is generated and released by urban traffic, factory production, and local cooling (the metabolism of the human body also generates heat, but it is usually only a small part of the surface energy balance).
- The urban surface stores more heat due to its geometry, materials, and reduced albedo, enhancing thermal retention.

- Artificial heat sources, including heating and cooling systems, contribute additional heat to the urban environment.
- The reduced vegetation and water bodies in urban areas limit natural cooling processes such as evapotranspiration.

By understanding these factors, researchers can better address the urban heat island effect and develop strategies to mitigate its impact.

As shown in Table 1.6, the energy balance framework explains the cause of the heat island effect by detailing how heat is transferred to and dissipated from the surface, adhering to the first law of thermodynamics, i.e., the law of conservation of energy. For the surface, energy conservation means that all energy either comes from solar radiation or anthropogenic heat dissipation; this heat is either used to warm the air, evaporates with water vapor, or is stored.

On a relatively macro scale, scholars have explored the correlation between function and the heat island effect, suggesting that the anthropogenic heat generated by urban functions is a significant cause. Land use directly reflects human activity, and research has shown that changes in activities within land use are the main reason for the increase in anthropogenic heat. This is not absolutely related to the area of urban construction land, indirectly proving that the type of function is one of the important factors affecting the heat island effect.

Previous studies found that among the four heat sources—industry, transportation, construction, and human metabolism—the contribution rate of industry is dominant, reaching 75.1%. The contribution rates of the other three types are 12.5%, 9.9%, and 2.5%, respectively. Some scholars also believe that traffic heat dissipation is a very important part of anthropogenic heat dissipation. Hart et al. [49] using Portland, USA as an example, derived a decision tree regression model for the intensity of the urban heat island during the daytime high-temperature period in summer. The results show that road closure and total road length on weekdays are the most important factors for the thermal environment, while the total road length on non-working days is not the most important factor. It is speculated that road traffic emissions

Table 1.6 The core characteristics of urban and suburban areas affecting heat island effect and urban surface energy balance

Key features that promote heat-island formation	Effects on the energy balance
Lack of vegetation	Reduce evaporation
Widely used imperpermeable surface	Reduce evaporation
Urban construction materials that are difficult to dissipate heat have increased	Increase heat capacity
The solar reflectivity of urban Construction materials is low	Increase net radiation
Urban geometry captures heat	Increase net radiation
Urban geometry decreases the wind speed	Reduce convection
Air pollution worsens	Increase net radiation
Increased energy adaptation	Increase human heat discharge

have a warming effect on the microclimate environment. Khan found in the statistics of anthropogenic heat in Brisbane, Australia, that during the peak period of heat dissipation in summer, traffic heat dissipation accounted for about 70%. Dhakal et al. used the DOE-2 software to calculate the anthropogenic heat of office buildings, commercial buildings, and residential buildings in the Tokyo area. They found that the city center, which is concentrated with commercial, administrative, and business complexes, has the largest amount of anthropogenic heat. They also found that during the day, the anthropogenic heat source is mainly office buildings, at night it is mainly commercial buildings, and in the late night and early morning, it is mainly residential buildings.

In summary, the causes of the formation of the heat island effect can be summarized as follows:

Reduction in Evaporation

Oke calculated the net radiation, convective heat dissipation, evaporation, and storage capacity for Vancouver, Chicago, and Mexico City during the day. He found that the most significant difference in the urban–rural comparison data was in evaporation, closely related to changes in vegetation cover (Table 1.7). Without evaporative heat dissipation, cities and suburbs store more heat during the day and release it at night.

Increase in Heat Capacity

Building materials generally have two properties that can exacerbate the heat island effect: thermal conductivity and heat capacity. Materials with good thermal conductivity can easily conduct heat into the material, while materials with high heat capacity can store more heat. The combination of these two properties is known as thermal diffusivity. A higher thermal diffusivity means that heat can be stored in the material, maintaining a relatively high temperature of the object for a longer time. The thermal diffusivity of man-made materials such as paving, insulation, and urban construction materials is much higher than that of natural materials like soil.

Increase in Net Radiation

Urban environments capture more net radiation than rural areas due to several factors, including the lower reflectivity of sunlight by urban materials, the unique geometric structures of urban areas, and air pollution.

Table 1.7 Land use, weather, and daytime energy balance in rural Vancouver, suburban Chicago, and built-up areas of Mexico City

		Rural Vancouver	Suburban Chicago	Built-up areas of Mexico City
Surface cover	The proportion of vegetation and water area	100%	44%	2%
	The proportion of building area	0%	33%	43%
	The proportion of the impermeable pavement area	0%	23%	55%
Weather	For precipitation during the study period	√	√	×
	Solar radiation (W/ m^2/Day)	5380	7000	4920
Daytime energy balance rate	Net radiation/solar radiation	0.66	0.60	0.15
	Controtive heat exchange/solar radiation	0.20	0.21	0.23
	Evaporative heat dissipation/solar radiation	0.45	0.28	0.03
	Energy storage/ solar radiation	0.04	0.19	0.31
	Energy expenditure/ solar radiation	0.03	0.08	0.42

Reduction in Convective Ventilation

Studies have found that the heat island effect is more intense in calm, clear weather because less heat is transferred from the ground to the air when wind speeds are low. Low wind speeds increase the amount of heat stored during the day and slow down the release of heat at night. In urban areas, buildings block the wind, resulting in a 60% decrease in wind speed, which makes urban ventilation less effective than in rural areas.

Increase in Anthropogenic Heat Emission

Anthropogenic heat is generated by human activities, including building cooling, vehicle emissions, and industrial production. To quantify the amount of anthropogenic heat produced in each area, all types of energy use must be considered, including commercial, residential, industrial, and transportation sectors. Typically, anthropogenic heat is higher in winter than in summer, with estimates of 20–40 W/m^2 in summer and up to 210 W/m^2 in winter [50]. Research indicates that anthropogenic heat has become more pronounced in recent years, mainly due to increased energy consumption, particularly the use of air conditioning during the summer.

1.5.4 Research on the Types of Urban Form Indicators Affecting Urban Heat Islands

The correlation between urban form elements and the heat island effect has always been a focus of scholarly research. Existing literature identifies potential urban form elements affecting the heat island effect, which can be summarized as functional development, surface coverage, construction intensity, and building combination.

1.5.4.1 Functional Development Indicators

The impact of functional elements on the urban heat island (UHI) effect can be categorized into direct and indirect impacts. The direct impact involves the external heat dissipation of different functional elements, particularly related to air conditioning systems in building groups. Chen et al. [12] noted that the method and location of air conditioning heat dissipation significantly affect the urban microclimate. The indirect impact mechanism is more complex and generally pertains to how urban functional layouts influence traffic flow. Vehicle exhaust not only directly generates heat emissions but also increases the air's absorption efficiency of solar radiation through greenhouse gases, thereby intensifying the UHI effect.

Previous studies have investigated the impact of urban functional elements on the UHI effect using two primary scales: block function and dominant functional area. These two scales differ significantly. Block function research focuses on a smaller scale, distinguishing functional differences in planning or current land use at the control level, with typically clear boundaries. Dominant functional areas, on the other hand, refer to city zones where specific types of urban land use are highly concentrated in relatively small areas.

Block Function

According to Kevin (1981) definition of urban form, urban function is a crucial element in shaping urban form. Scholars have used urban functional elements as a starting point to explain the urban heat island (UHI) effect, finding that different functional types have varying impacts on the UHI.

However, existing studies lack detailed and in-depth research on the relationship between land use in built-up areas and land surface temperature. At the overall urban level, some research employs remote sensing data to make rough divisions of construction land. Other studies use urban planning land use maps and land surface temperature inversion results to determine that commercial and industrial land have higher temperatures, while park green spaces have lower temperatures. For instance, Li et al. [51] identified five types of functional land in Shanghai—new residential areas, old residential areas, villas, industries, and public facilities—and compared the average surface temperature differences and variances among them. Song [52] measured the land surface temperature of different functional lands in Bundang town and found that residential and commercial areas were 4.5 °C warmer than forests, followed by transportation facilities. Qihao Weng et al. [53] concluded that industrial land temperatures were higher than those of residential and commercial areas, even when these areas were high-density. Previous studies observed the frequency and intensity of heat islands in commercial areas, industrial areas, communities, and forest parks, finding that industrial areas and forest parks had a greater positive impact on heat island frequency and intensity. Commercial areas generally had a positive effect on the heat island effect, while forest parks tended to have a negative effect.

Overall, related studies have found that industrial, commercial, public service facilities, transportation facilities, and high-density residential areas with high built-up density and a high proportion of impervious surfaces exhibit higher surface temperatures. Previous studies found in a study of Berlin that the land surface temperature varied greatly depending on land use when functional land types were typical. Those samples with insignificant differences can be explained by the following disturbing variables: (1) the difference in size; ② the inhomogeneity of functional land use; ③ Some functions have greater similarity to each other (such as mixed land and residential land).

This type of research employs a comprehensive quantitative method, regarding the entire urban area as the research object. After identifying specific functional lands, it integrates their relationships with the heat island effect. However, the land analyzed often represents only a part of the overall built-up area, and the extracted land is fragmented. Consequently, the land surface temperature of the analysis object may be significantly affected by the surrounding environment, making it difficult to objectively present the thermal environmental characteristics of that specific urban function.

Dominant Functional Area

In addition to studying the functions of subdivided land use at a high resolution, analyzing the correlation between land use and the heat island effect based on several major categories of dominant functional zones can more intuitively express the heat island effect of different urban functional areas. This approach can also help avoid the interference of external effects on the analysis results. Jusuf et al. [54] categorized the dominant functional areas along the south coast of Singapore from west to east as industrial, commercial, residential, and airport areas. They combined this categorization with east–west cross-sectional land surface temperature data to analyze differences in average heat island intensity and the spatial distribution characteristics of different functional areas. For example, residential areas showed scattered high-temperature zones due to the presence of green open spaces, whereas industrial and airport areas, which have large, concentrated hardened surfaces, exhibited more concentrated high-temperature areas.

1.5.4.2 Land Surface Cover

Landscape of Land Cover from a Geographical Perspective at the Macro Scale

At the macro level of urban form analysis, scholars often use remote sensing data to obtain surface cover conditions [39, 53, 55, 56]. For convenience in data acquisition, researchers employ commercial software for various classification techniques, such as supervised and unsupervised classification, target recognition, and decision tree classification (e.g., ENVI, ERDAS). These methods result in discrete landscape classifications, identifying the main types of surface cover for each pixel, such as buildings, impervious surfaces, grasslands, forests, and water bodies [57], and allow for the analysis of surface temperature across these different cover types.

Li [58] proposed that urban built-up areas have a double-layer structure comprising a surface layer of artificial materials (concrete, asphalt) and a sub-surface matrix layer of natural materials (clay, sand, pebbles). This unique structure facilitates efficient heat storage, characterized by low surface reflectance and insufficient evaporation. Additionally, the vertical walls of the city and the "canyons" formed by these structures not only reduce horizontal radiation but also enhance heat exchange efficiency. Due to these factors, built-up areas are significant spatial elements in the study of the heat island effect.

Research on the relationship between artificial land cover and the heat island effect has been conducted across different cities. Previous studies investigated the overall built-up area of cities by analyzing morphological characteristics, creating a sample library from multiple cities, and comparing the heat island effects across these cities. Through land surface temperature inversion, Zhang identified heat island zones in ten cities in the Pearl River Delta region and found a significant positive correlation between the area of built-up development and the extent of heat island effects. Similarly, Ding Jincai observed that from 1992 to 1998, the built-up area in Shanghai doubled, leading to a corresponding doubling of the heat island effect, highlighting a close relationship between built-up area expansion and increased heat island intensity. Liu et al. [31] also found a significant positive correlation between the built-up area and the heat island effect in Beijing, noting that both the total built-up area and its increase were significantly associated with heat island intensity. This pattern is also evident in European cities, where Schwarz and Manceur [37] reviewed the relationship between the built-up area and the heat island effect. Schwarz found

Table 1.8 Landscape metrics used to measure the correlation between built-up area and heat island effect

Name of index	Description
Percent of cover	Area proportion of the built-up patch
Mean patch area	Total patch area divided by the number of patches
Patch density	Total patch area divided by total area
Edge density	The length of all patch edges per hectare
Landscape shape index	Total patch side length divided by the length of the largest block
Mean patch shape index	/
Mean euclidian nearest-neighbor distance	The average closest distance between patch edges

that the area of built-up land is proportional to the average temperature of the area and that compact built-up areas are more likely to exacerbate the heat island effect compared to dispersed ones.

Land Cover Research from the Perspective of Landscape Ecology at the Mesoscale

The size and shape of urban built-up areas significantly influence land surface temperature, especially within the scale range of 100 m^2–1 km^2 [59]. At this mesoscale, the land cover exhibits complex landscape patterns, necessitating an analysis of the relationships between different surface cover patches. Schwarz and Manceur [37] examined various landscape pattern indices, including the number of spatial patches, average patch size, median patch size, edge density, area-weighted spatial fractal dimension, and area-weighted average shape index. Schwarz verified the correlation between these indices and the average heat island effect in cities. Previous studies analyzed the relationship between urban built-up area morphology and the heat island effect in Tianjin from 1992 to 2013. Using eight urban form indicators—area, density, edge characteristics, shape, core area, independence, proximity, contrast, sprawl, dispersion, connectivity, and diversity—Huang identified that dispersion, proximity, fractal dimension, and the patch area-to-perimeter ratio are most closely related to heat island intensity. Weng et al. [60] found that the complexity of residential area plans is associated with greater variability in surface temperature, with more complex residential designs resulting in greater temperature fluctuations. Zhou et al. [61] used high-resolution remote sensing data from Baltimore to demonstrate that built-up area surface temperature correlates significantly with multiple landscape pattern indicators, including maximum patch index, patch density, edge density, average patch size, standard deviation of patch scale, and average shape index (Table 1.8).

To more precisely uncover the correlation between built-up area landscape patterns and the heat island effect, some scholars have shifted the research scale from city-wide analyses to more localized street-scale investigations. Connors, Deng et al. [62] studied the impact of landscape patterns on the heat island effect by selecting 90 sample areas in Phoenix, Arizona, USA, comprising 30 samples each from industrial, dry, and wet zones. They analyzed the relationship between landscape pattern indices—such as surface cover proportions, patch density, edge density, shape index, patch number, fractal dimension, sprawl, and diversity—and the heat island effect. Following this approach, Rotem-Mindali et al. [63] focused on the proportion of surface cover types rather than landscape patterns. They selected five sample types from high-green residential areas, general residential areas, urban parks, and industrial zones to study the relationship between building cover proportion, normalized difference vegetation index (NDVI), and surface temperature. Their findings indicated that land surface temperature is generally positively correlated with building cover and inversely correlated with NDVI. However, some researchers question the effectiveness of these landscape pattern indices for heat island effect studies. Chen et al. [64] argued that landscape pattern indices need careful selection and that two-dimensional landscape patterns offer limited insight into the heat island effect. Factors such as floor area ratio, building height, and shadow effects play more significant roles.

Land Cover Research

Recent studies have increasingly focused on how the composition of surface cover influences land surface temperature. The role of vegetation and water bodies in mitigating the heat island effect has become a prominent research topic [65–67]. Empirical studies suggest that green spaces and water bodies help reduce heat island intensity through daily transpiration, CO_2 absorption, and dust trapping [53, 61, 68, 69]. Despite these benefits, large-scale green space implementation in big cities remains challenging (Gill et al. 2007). Many studies use the Normalized Difference Vegetation Index (NDVI) to quantify green space coverage and its relationship with the heat island effect, demonstrating a significant negative correlation between vegetation and urban surface temperatures [70]. Additionally, research by Stabler et al. [71], Chen et al. [30], and Seungbum [72] confirms this negative correlation.

Scholars have also explored how vegetation configuration and coverage impact the heat island effect. Previous studies compared different green space types in Changsha, finding that tree forests have the greatest cooling effect, followed by tree-shrub forests, shrubs, and grasslands. Weng et al. [53] examined vegetation coverage in Indianapolis, finding that vegetation coverage correlates better with surface temperature than NDVI. Their study showed that various land use types and spatial patterns affect this relationship.

To better guide urban planning through quantitative analysis of ecological resources' impacts on the heat island effect, researchers have studied green space and water systems, focusing on scale and form as key variables. Previous studies investigated how park green space area, perimeter, and shape index affect thermal environment effects, emphasizing that scale is a crucial factor in future research. Yue

(2008) studied the heat island effect in Shanghai, establishing a 250 m buffer around urban park patches to measure the cooling effect. Yue found that the cooling effect of park patches is more strongly related to their shape complexity rather than their size or length. The study also compared the cooling effects of the Huangpu River, Suzhou River, and Chuanyang River, concluding that wider rivers closer to the heat island center have a stronger cooling effect on the urban thermal environment.

Green spaces and water bodies differ in their cooling effects. Water bodies, due to their relatively stable temperature, are particularly effective at cooling the city center where the heat island effect is most severe. However, the effectiveness of green spaces in mitigating the heat island effect remains debated, with some studies showing that the shape and density of green spaces do not always follow a predictable pattern [73]. Researchers argued that the relationship between green space form and the heat island effect exhibits spatial scale sensitivity. Huang's research used varying buffer distances to calculate the sensitivity of green space density to heat island intensity, finding that a 15 m radius provides the greatest impact on reducing heat island effects.

1.5.4.3 Construction Intensity

Existing literature on the correlation between construction intensity and heat island intensity is relatively sparse, and often fails to distinguish clearly between "construction intensity" and "building combination." In urban planning, construction intensity is typically measured using indicators such as building density and floor area ratio. Building density refers to the concentration of buildings within a given area, while the floor area ratio indicates the total building area relative to the land area. Conversely, building combination focuses on the arrangement and structural configuration of building groups.

Most research indicates that regions with higher construction intensity contribute to increased local heat capacity due to the substantial volume of building materials, which impedes the rate of heat dissipation. Building density, as a spatial morphological indicator, effectively characterizes heat island intensity. Higher building density can reduce wind speed, which in turn affects the heat island effect. Building density influences the heat island effect by modifying local wind patterns and heat distribution. Ng et al. [74] developed a correlation model demonstrating that increased building density correlates with reduced wind speed, exacerbating the heat island effect. Buildings with higher floor area ratios obstruct the outward scattering of long-wave radiation, thereby slowing the "leakage" of radiant energy into the sky and resulting in higher urban temperatures compared to suburban areas [75].

Overall, while construction intensity and building combination are interconnected aspects of urban form, they have distinct impacts on the heat island effect. Building density primarily affects local microclimates through wind and radiation interactions, whereas building combination deals with the spatial arrangement and structural characteristics of urban environments. Understanding these differences is crucial for developing effective urban planning strategies aimed at mitigating heat island effects.

1.5.4.4 Building Morphology

Relevant research highlights the building group as a significant component of the urban underlying surface and recognizes changes in the three-dimensional shape of this surface. A review of the literature reveals that urban spatial morphology can be quantified using several key indicators, which can be categorized into the following dimensions:

- Spatial Distribution Elements: These indicators assess the relative concentration or dispersion of objects at a given scale compared to a larger-scale reference. They are used to quantify aspects such as the dispersion and uniformity of spatial distribution.
- Proximity Elements: Proximity measures the average distance between two objects, focusing on the spatial relationships and separations between different elements.
- Connectivity Elements: This category evaluates the relative accessibility and spatial interconnectivity between different patterns, highlighting how elements are connected or separated.
- Diversity Elements: Diversity refers to the variety and mix of objects within a given area, such as variations in land use or building scales. It examines the proportion of different types of objects rather than their spatial aggregation.
- Morphological Elements: These elements describe the physical characteristics of objects, including their shape, volume, and the space they occupy.

Overall, scholars utilize these spatial morphological elements to investigate the impact of building combinations on the heat island effect through two primary dimensions: convective heat transfer and radiant heat effects. Convective heat transfer refers to the movement of heat through air or other fluids, while radiant heat effects involve the absorption and emission of heat from surfaces. By analyzing these dimensions, researchers aim to understand how different spatial arrangements and building configurations contribute to urban heat islands.

Wind Environment and Convective Heat Transfer

The artificialization of urban surfaces significantly impacts the wind environment and convective heat transfer. Different building group characteristics influence the ventilation rate in varied ways. Researchers have utilized meteorological scale software, such as WRF (Weather Research and Forecasting) and MM5 (Fifth-Generation Penn State/NCAR Mesoscale Model), to simulate these relationships on a macro scale across entire cities [76]. These studies comprehensively consider factors such as urban form and building configurations, positing that the shape of the city and ground features are crucial determinants of the urban wind environment.

Urban shape parameters commonly studied include the building density (coverage ratio of the building plan area on the grid), the density of the urban canopy area (describing the extent to which buildings occupy the airspace), and the drag coefficient (which reflects the number, spatial distribution, height, and shape of buildings, as well as the angle between airflow and buildings). Surface feature parameters include air dynamic roughness and zero plane displacement. By incorporating artificial heat flux and solar radiation into these models, researchers have refined their simulations of the urban heat island effect. Initially developed for improving local weather forecasts, these models have been advanced to depict urban spatial morphology in greater detail [77], employing large-scale three-dimensional urban models to analyze single-factor variables affecting the urban thermal environment and compare different planning schemes [51].

However, studies using regression models to identify patterns face two main challenges. First, the three-dimensional configuration of building groups presents difficulties for fully and accurately capturing all morphological attributes using existing indicators. Second, controlling for irrelevant variables and external influences in the analysis of specific urban form indicators and their effects on the heat island phenomenon remains a significant challenge. Consequently, there has been increasing scholarly attention towards advanced modeling techniques, such as Computational Fluid Dynamics (CFD), to establish turbulence models that simulate temperature fields, velocity fields, and concentration fields resulting from fluid flow within urban spaces (Yu and Zhang 2007). CFD-based research focuses on detailed descriptions of complex geometric boundaries and flow fields, revealing the underlying mechanisms of atmospheric movement and microclimate conditions on a smaller scale (Wang 2008). Notably, some researchers have employed CFD models to quantitatively analyze the effects of various influencing factors and building layout forms on the urban thermal environment through case studies of different building group configurations, such as semi-enclosed, fully enclosed, row, and staggered arrangements, as well as the ratio of building spacing to building height. These studies provide a theoretical foundation for predicting and evaluating urban microclimates.

Solar Light and Radiant Heat

The urban form of urban blocks, the shape of buildings, and the nature of block interfaces are key morphological elements that influence the urban microclimate at the building group level. These factors affect both the ventilation and solar radiation absorption capabilities of the building group, as well as the heat storage capacity of the underlying surface [12]. Research has examined the impact of building group morphology on heat dissipation potential from the perspective of heat transfer, identifying several important parameters such as the shape coefficient, shadow density, and sky view factor.

a. Shape Coefficient

The shape coefficient, which represents the ratio of the building's surface area to its volume, describes how a building interacts with the external environment through natural ventilation and solar radiation. Buildings with a high shape coefficient, such as those found in compact blocks with multiple courtyards, absorb substantial amounts of heat during the day and release it at night. This characteristic enhances the thermal inertia of the building group, making it an important factor in the urban heat island effect. For example, in a built-up area with complex morphology, heat absorbed during the day is retained and subsequently released at night, thereby affecting the local microclimate.

b. Shadow Density

Shadow density refers to the amount of shaded area on the ground observed at fixed time intervals. This parameter is highest in streets with courtyard structures and narrow lanes, which create internal microclimates by providing shaded areas. In contrast, modern standalone buildings typically lack the ability to create such microclimates due to their different spatial configurations. By influencing the distribution of shadows, shadow density affects how solar radiation is absorbed and dissipated in urban environments.

c. Sky View Factor

The sky view factor, a critical parameter in urban climatology, quantifies the extent to which the sky is visible from a specific location. According to Oke's formula, the relationship between the maximum temperature difference (ΔT_max urban–rural) and the sky view factor (a) is expressed as:

$$\Delta T_{maxurban-rural} = 15.27 - 13.88a$$

where ΔT_max urban–rural represents the maximum temperature difference between urban and rural areas, and SVF denotes the sky view factor. Traditional courtyard buildings have a larger sky view factor compared to modern standalone buildings, which helps to moderate nighttime temperature drops and mitigate rapid temperature increases during the day due to direct sunlight.

ZED (1997) studied six common urban forms in the context of London's climate and concluded that courtyard structures are the most effective for capturing solar radiation. Similarly, previous studies summarized architectural group morphological factors affecting the climate into several key indicators, which include: ① volume-surface area ratio; ② passive space; ③ changes in surface area and horizontal section perimeter with height; ④ longitudinal section and orientation; ⑤ the impact of design complexity. These factors elucidate how various morphological characteristics of buildings influence local climate conditions and the heat island effect.

1.5.5 Urban Heat Island Effect Prediction and Urban Form Optimization Research

1.5.5.1 Heat Island Effect Warning

Urban Climate Map (UC map) is a widely accepted and mature tool for predicting the heat island effect on a macro scale. Initially designed for analyzing and evaluating urban climate environmental information, UC map utilizes two-dimensional spatial analysis to assess climate phenomena and address existing air pollution and heat island issues [78, 79]. Over time, cities such as Tokyo and Hong Kong have adopted UC map techniques to tackle their specific climate challenges related to the heat island effect.

Early research in climate environment zoning largely relied on expert judgment and subjective assessments, as exemplified by the climate zoning approaches employed in Tokyo, Stuttgart, and other cities. However, with the advancement of remote sensing technology, more quantitative methods have emerged for climate simulation. For instance, Swiss researchers utilized remote sensing satellites to collect data on urban surface physical properties and land use, thereby automating the generation of climate environment zoning through the analysis of daytime thermal environments and surface roughness [80]. Additionally, FITNAH (Flow over Irregular Terrain with Natural and Anthropogenic Heat Sources) models have been used to simulate Berlin's thermal environment, revealing a high degree of consistency between simulated and actual conditions [81, 82]. In their study of Hong Kong, scholars argued that relying solely on land use and surface coverage was inadequate for capturing the physical characteristics of high-density settlements. They proposed incorporating more detailed urban form information, such as towers, plateaus, streets, and open spaces, and used metrics like building area and building density to quantitatively assess urban form [83].

Historical research on climate environment zoning demonstrates that while the built environment significantly affects the thermal environment, its impact on the wind environment is more limited. Instead, natural ecological spaces play a crucial role in shaping wind conditions. At the macro scale, the natural environment is the primary determinant of wind conditions, while the built environment acts as a disruptive factor. Consequently, research in this area often focuses on analyzing large-scale ecological factors, such as terrain, water bodies, and natural landscapes, and their effects on local atmospheric circulation, prevailing wind directions, and potential wind channels. In contrast, at the micro scale, the built environment's influence on the thermal environment is more pronounced, with urban design elements such as building height, volume ratio, and surface reflectivity being critical factors.

To address these micro-scale interactions, scholars have developed the Local Climate Zone (LCZ) theory (Table 1.9). The LCZ concept offers a more intuitive understanding of temperature variations within the city and provides a systematic framework for characterizing urban physical environments in relation to the heat island effect. The LCZ theory includes eight morphological indicators and a set of quantitative classification criteria for identifying different LCZ types in a research area. These indicators are:

- Sky View Factor
- Surface Reflectivity (Aspect Ratio)
- Building Density
- Proportion of Impervious Surfaces
- Proportion of Permeable Surfaces
- Roughness of Building Groups
- Terrain Roughness.

The LCZ theory has garnered significant attention in the academic community, leading to a series of quantitative validations and refinements of the original classification system. This theoretical framework helps researchers explore the correlation between urban forms and the heat island effect from a unified perspective, offering innovative contributions to micro-level urban climate (Table 1.9).

1.5.5.2 Numerical Simulation of the Thermal Environment

For an extended period, numerical models have been highly valued for studying the impact of urban spatial structure on the atmospheric environment. From the perspective of urban climatology, the proliferation of high and low buildings has increased the roughness of the urban underlying surface, altering atmospheric dynamics [73] and creating a significantly different atmospheric boundary layer in urban areas compared to the flat and uniform layers found in suburban regions. In the special near-surface layer of the urban atmospheric boundary layer, the underlying surface structure and human activities directly affect the thermal and dynamic processes of the urban atmosphere [84]. The advancement of computational technology has enabled more sophisticated numerical simulations of the heat island effect, allowing researchers to explore this phenomenon in the absence of direct temperature measurement points. Since numerical simulations were first used to model the heat island effect, the field has evolved from simple one-dimensional models to complex three-dimensional models. Kimura and Takahashi [85] employed a three-dimensional dynamic model to study the urban heat island effect in Tokyo. Their comparison with actual measurements and sensitivity tests demonstrated the model's credibility and revealed the effects of artificial heat discharge and green space on the heat island phenomenon.

In recent years, the development of urban canopy models has led to their integration with mesoscale numerical models such as MM5 and WRF, yielding promising simulation results. Historically, models based on MM5 and WRF were constructed for flat and homogeneous surfaces, relying on traditional Obukhov near-surface similarity theory and energy balance theory for simulations. These models sought to identify dynamic and thermal characteristics different from other surfaces to reflect the urban environment's impact on atmospheric dynamics and thermal structure while overlooking the heterogeneity introduced by urban forms. Some scholars have

Table 1.9 Classification criteria for local climate zone

LCZ type	Definition	Building height	Building density	Diagrammatic sketch
LCZ 1 (high density, high height type)	Dense buildings of 10 stories and above; fewer or no trees; most of the surface is covered by pavement; building materials are mainly cement, steel, glass, etc	> 30 m	40–60%	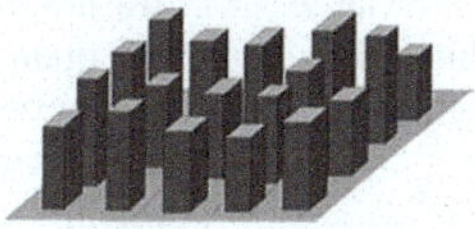
LCZ 2 (high-density, medium-height type)	Dense 3–9 storey buildings; few or no trees; most of the surface is covered by pavement; building materials are mainly brick, cement, etc	9–30 m	40–70%	
LCZ 4 (low density, high height type)	10 stories and above buildings with open layout; most of the surface is covered by permeable material; the building materials are mainly cement, steel, glass, etc	> 30 m	20–40%	
LCZ 5 (low-density, medium-height type)	3–9 floors; most of the surface is covered by permeable material; building materials are mainly brick, cement, glass, etc	9–30 m	20–40%	
LCZ 6 (low density, low height)	1–3 floors; abundant permeable surface; building materials: wood, brick and concrete	3–10 m	20–40%	
LCZ 7 (light weight, low height)	Dense single-story buildings; few or no trees; surface covered mainly hard surface; lightweight building material	2–4 m	60–90%	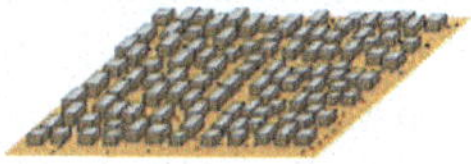

(continued)

Table 1.9 (continued)

LCZ type	Definition	Building height	Building density	Diagrammatic sketch
LCZ 8 (large volume, low height)	Low height buildings with open layout; small amount of goods without trees; pavement basically covering the surface; building materials are stone, steel and concrete	3–10 m	30–50%	
LCZ 9 (scattered building complex)	Small and medium sized buildings; large permeable surface	3–10 m	10–20%	
LCZ 10 (heavy industry zone)	Medium and low height industrial buildings; few or no trees; pavement covered with base table; building materials of metal, concrete	5–15 m	20–30%	

addressed this issue by improving the accuracy of mesoscale models in depicting urban surface forms. For example, Brown, Masson, and others proposed more detailed parameterization schemes for urban surfaces [86, 87]. However, these schemes primarily describe idealized urban structures.

Li et al. [88] advanced this field by designing the Urban Canopy Parameterization (UCP) scheme, which offers a more comprehensive and realistic representation of urban atmospheric interactions. This scheme divides the urban canopy into vertical layers, differentiating architectural structures at various height intervals. The parameters used include building coverage ratio, urban canopy area density, drag coefficient, and average building height, which are integrated with atmospheric thermodynamics and dynamic equations to simulate the heat island effect. Following this, scholars have parameterized urban canopy forms from various perspectives. Chen et al. [30] conducted sensitivity tests on different building heights and densities, demonstrating the significant impact of building configurations on urban airflow and boundary layer structure. Wang [39] developed a multi-layer urban canopy model incorporating three-dimensional building distributions and architectural features like building width, spacing, and vertical density to better reflect urban surface impacts on temperature, humidity, and momentum equations.

Despite these advancements, simulation research on the urban heat island effect often faces the challenge of coarse granularity, with the smallest grid size typically being 500 m × 500 m. This granularity fails to capture the intricate spatial textures of urban environments and does not account for the effects of specific urban form parameters on the heat island effect. To address this, some scholars have performed

Table 1.10 Comparison of influencing factors on urban heat island

Factor	Conventional scheme	Recommended to control the planning indicators
Setting and optimization of the ventilation corridor	Did not consider	Set up ventilation corridors
Building energy saving measures	Did not consider	No less than 20% more energy saving than the conventional scheme
Roof greening	Did not consider	No less than 30% of residential plots; no less than 42% of public plots
Exterior wall reflectivity	0.1	Of 0.3 or more
Ratio of green space	30%	Not less than 50%

diachronic analyses to examine the relationships between key building group parameters and the heat island effect on a smaller scale. This includes exploring factors such as floor area ratio, building density, green space ratio, water surface ratio, and hard surface ratio, as well as deriving sensitive scale radii and spatiotemporal characteristics for heat island formation. Previous studies developed an urban heat island simulation prediction and warning information system using C# language, which enables spatial simulation of heat island intensity, warns of heat islands based on floor area ratios and hardened ground, and guides the spatial optimization of green spaces and water bodies.

This pioneering work in heat island prediction based on urban form has been furthered by Liu et al. [89], who utilized numerical methods to explore the energy and material transfer relationships between building internal climates and local urban climates. They developed a local urban-building coupled climate evaluation model that includes a local climate module, a building thermal load calculation module, a solar radiation module, and a thermal comfort module. This model simplifies the distribution of building communities and has been validated for dynamic quantitative evaluations of urban thermal climates through comparisons with measured data. Researchers extended this model into practical urban planning applications by translating model analysis results into detailed urban land use regulation indicators (Table 1.10).

1.5.5.3 Urban Form Optimization Strategy

Many municipalities employed a common strategy to create an effective urban ventilation environment, which is examining the planning practices aimed at reducing heat island intensity in cities globally and optimizing ecological spaces and architectural forms.

Germany has always been at the forefront of the world in the field of urban ventilation corridor planning research and practice. As early as the 1970s, climatologists began to participate in urban planning, Since the 1990s, the development of computer technology has enhanced the integration of urban climatology requirements into planning processes. German wind tunnel research suggests that the effectiveness of an urban ventilation system depends on two factors: the morphology and structure of the wind tunnel's underlying surface, and the climatic function of each component of the wind tunnel system and its spatial and functional connections. The material spatial characteristics of the underlying surface determine its climatic function, which can be divided into functional spaces, compensation spaces, and air guidance channels, typically referring to built-up areas, ecological areas, and ventilation corridors, respectively. To fully utilize local circulation for building an urban ventilation system, it is essential to clarify the planning and control strategies for these three types of underlying surfaces: ① maintain and expand the climate regulation capacity of compensation spaces, especially protecting areas that positively affect local circulation; ② build ventilation corridors and establish effective connections between compensation spaces and functional spaces.

Based on the foundation of German wind tunnel control principles, Japan and Hong Kong's wind tunnel planning has further refined the morphological index regulation system to meet their unique geographical condition. Tokyo's "Sea Forest" wind tunnel planning in Japan identified four key areas for wind environment improvement design based on a thermal environment analysis map: high-density commercial areas, high-density residential areas, and future development areas. For each area, detailed planning and control guidelines were established. The main transformation measures include increasing open spaces and green spaces and reserving important ventilation corridors. Hong Kong integrates land use, existing buildings, and other data to develop strategic suggestions for improving the urban climate environment for each climate planning unit. These strategies primarily involve strengthening existing wind tunnels, establishing new wind tunnels, pushing buildings back from narrow streets, and preventing further increases in building volume and impermeable surface coverage [90].

Subsequently, mainland Chinese cities such as Beijing, Fuzhou, Changsha, Shenzhen, and Wuhan have successively proposed wind tunnel planning (Table 1.11) to optimize the overall ventilation pattern by regulating urban spatial morphology. The indicators involved are primarily related to planning structure, building groups, single buildings, and green space layout, with a particular emphasis on the control of building layout adjacent to ecological areas.

1.5.6 Comprehensive Review of Related Research

1.5.6.1 Urban Spatial Morphology

Past research has focused on the representation of material space in urban spatial morphology, typically grounded in human subjective experiences. These researches emphasized aesthetic and cognitive principles inherent in human consciousness.

Table 1.11 A list of building morphological control indexes of wind corridor pattern in various cities in China

Guiding level	Guiding morphological elements	Explanation
Planning structure	Corridor-ventilation system	The ventilation ducts should be connected to the urban ventilation corridors to form a continuous Corridor-ventilation system, effectively channeling wind into the urban fabric. The angle between the duct and the ventilation corridor should ideally be around 90°. Attention should be paid to avoid obstacles protruding into the ventilation corridor-duct system
	Corridor form	When urban space such as streets is used as ventilation corridors, the ventilation effect of concentrated and wide main streets is better than that of scattered and narrow streets
	Main road angle	In order to improve the average wind speed in the region and ensure that each individual building can obtain sufficient natural ventilation, it is appropriate to make the direction of the main road and the dominant summer wind direction at an Angle of about 30–60° in the high-density urban development area or the area with a plot size of more than 100 m, and make the long side of the plot parallel to this direction
	Height control	Buildings near the regional center of land and water wind resources such as mountains / rivers should not be too high
Building groups	Building density	The building density, building height, building ratio above 60 m and aspect ratio at the vents should be strictly controlled
	Building width	
	Building width and height ratio	
	Higher than 60 m of the building proportion	
	The aspect ratio of the building	
	The direction of the pedestrian street	Pedestrian street should use the north–south direction or close to the north–south direction
	Building types	It is advisable to use staggered, diagonal, or free-form arrangements that integrate with the terrain. The three-dimensional layout of the buildings should adopt a 'stepped' treatment, avoiding the formation of completely enclosed spaces. The height of the buildings should not be uniform. The height difference should be stepped; buildings closer to the dominant wind direction should be lower. If there is no clear dominant wind direction throughout the year within the site…

(continued)

Table 1.11 (continued)

Guiding level	Guiding morphological elements	Explanation
	Interstitial rate	Interstitial rate = building surface width / base surface width. The vents must be controlled with adequate open surfaces to facilitate air access into the ventilation corridor from the compensation space
Building monomer	Overhead rate requirements	Ensure that a certain proportion of the building ground floor is overhead
	Building orientation	In order to ensure the good ventilation quality of the group, it is advisable to form a 30° and 60° angle between the building and the long side dominant wind direction
	Building types	When the windward side of the building is long, it is appropriate to facilitate the wind guide The single building should avoid the ventilation dead angle formed by the completely enclosed space
Green cover	Green coverage ratio	The green coverage of the wind crossing should not be too low
	Green three dimensional quantity	
	Ratio of green space	

Scholars like Kevin Lynch and Schultz have moved beyond the traditional entity space research of classical urban design, which often prioritizes "seeing things but not people." They have incorporated environmental psychology to perceive and regulate urban spatial morphology, marking a significant advancement in the theory of urban spatial morphology.

However, in the context of contemporary urban responses to climate change, urban form theory grounded in aesthetics and psychological cognition struggles to guide the objective and physical goals of mitigating climate change. The theory of urban form needs the support of engineering science. As another spatial analysis perspective oriented towards physical space, the theories, and methods of engineering science in urban form research have surpassed the tradition of understanding urban form through subjective cognition, moving towards a more scientific and rational dimension. Specifically, with the concepts of sustainable development, low carbon, and ecology increasingly ingrained in people's minds, recent years have seen domestic and international research on urban physical space shift from subjective feelings to engineering science, aiming to reveal how different physical space features affect performance indicators such as carbon emissions, energy consumption, wind environment, and heat emissions. Particularly in the field of building physics, numerous studies have emerged on the correlation between the morphological elements of a single building or small building group and its performance indicators. The study of urban form in the field of building physics reflects the disciplinary trend of engineering science, which holds significant reference value for this research.

It is worth noting that existing literature has extensively studied the correlation between land cover and land surface temperature, but most of these studies utilize indicators derived from calculating the normalized vegetation index, normalized water body index, and normalized building index through remote sensing bands. They do not directly address the various types of land use regulated by statutory urban planning. This book posits that the ability of urban and rural planning and building physics to depict urban form and the physical environment is undoubtedly more accurate than remote sensing indicators, which is conducive to guiding urban planning practice.

1.5.6.2 Heat Island Measurement

The current methods for calculating the heat island effect primarily include using remote sensing data to calculate land surface temperature and utilizing air temperature data from fixed temperature points. The former measures the temperature of the fixed land surface, while the latter captures the temperature of the moving air. Land surface temperature is the temperature produced by ground materials themselves after being radiated by sunlight, whereas air temperature reflects the temperature of the air. Due to the ubiquitous presence of near-ground airflow, the air temperature at a certain point is significantly influenced by the surrounding environment.

Air temperature data and land surface temperature data each have their strengths and characteristics. Temperature measurement point data is known for its high spatiotemporal accuracy. In terms of measurement accuracy, air temperature data measured at specific points can directly reflect the impact of the urban heat environment on the human body. The air temperature data measured at temperature measuring points can directly reflect the influence of the urban thermal environment on the human body, and can also know the land use, land cover, building air conditioning heat emission, and other conditions around temperature measuring points, so it is more meaningful to study the heat island effect in a small scale. In comparison, remote sensing first estimates the impact of the atmosphere on surface heat radiation. By subtracting this atmospheric impact from the total heat radiation observed by the satellite sensor, the intensity of surface heat radiation is obtained, which is then converted into the corresponding surface temperature based on the self-attributes of surface materials.

In terms of spatiotemporal accuracy, the spatial coordinates of temperature measurement points are more precise, allowing for automatic temperature measurement at frequent intervals, which facilitates historical analysis. However, remote sensing data, typically in grid form with a resolution of 30–500 m, has inherent spatial misalignments when matching with specific spatial coordinates. Additionally, obtaining remote sensing data of a particular city is often hindered by cloud cover. Researchers usually need to select data sets with minimal cloud cover, limiting the time phases of available data and making historical research challenging.

Remote sensing data excels in its broader coverage. While temperature measurement points are typically limited in their urban distribution and capture the gas temperature within a restricted area around them, remote sensing data offers extensive coverage. Temperature measurement points, often placed by meteorological departments, tend to avoid built-up urban areas to obtain air temperature data free from artificial heat discharge, leading to a failure in capturing temperature variations across different urban areas. In contrast, satellite data sources such as Landsat 8 and QuickBird provide comprehensive coverage of surface temperature distribution, offering a significant advantage in terms of measurement range.

Considering the research purpose of this article and the difficulty of obtaining data, remote sensing-based surface temperature data will be more conducive to the study. While air temperature data can reveal detailed temperature changes over time and establish a connection with urban form, its limitations make it less suitable for high-precision modeling. Specifically, due to the small sample size, it is challenging to establish a highly accurate regression model between urban form and air temperature. Additionally, the fixed locations of air temperature measurement points raise concerns about whether these locations are representative of the broader urban environment or if they are influenced by irrelevant factors such as artificial heat discharge or ecological spaces, which can compromise the quality of the sample data. In contrast, remote sensing data offers a significant advantage through its ability to capture surface temperatures on a large scale. This broad coverage facilitates the comparison of heat island intensity across different regions and supports the investigation of the relationship between urban form and heat island effects. Therefore, this book intends to utilize land surface temperature data as the foundational basis for research on the heat island effect.

1.5.6.3 Research Gap of Mechanism Explanation

Focus on Individual Attributes, Light on Environmental Impact

From a micro perspective, the heat island effect is influenced not only by the material properties and human heat discharge within a local area but also by the surrounding environment. Specifically, the thermal environment of urban local areas is significantly affected by the upwind area's conditions due to the fluid nature of gases, which facilitates the transfer of heat and pollutants from adjacent regions [91]. However, past studies, whether based on temperature point data or remote sensing data, have generally focused on attributing such thermal variations solely to the internal physical characteristics of the samples themselves. For instance, the LCZ (Local Climate Zone) theory classifies urban local areas based on material urban form elements and assumes that different classifications impact the thermal environment differently based on predefined index ranges. While this theory provides a framework for understanding the thermal environment within specific zones, it overlooks the influence of external factors and the location of the sample. As a result, it fails to explain why similar LCZ classifications can exhibit significant temperature differences across different areas, or why LCZs with varied characteristics can show similar temperature patterns in different locations.

Similar studies often overlook the interactive influence between thermal environments, focusing predominantly on "internal factors" while neglecting "external factors." According to dialectical materialism, changes in phenomena result from the combined effects of both internal and external factors, where internal factors serve as the basis for change and external factors provide the conditions for change, working through internal factors. Applying this perspective to the thermal environment of a local area, it becomes evident that while internal factors such as building configurations, land use functions, and surface coverage determine the radiant heat characteristics of the area, the surrounding environment plays a crucial role through convective heat transfer. Therefore, a comprehensive understanding of the thermal environment must consider both internal attributes and external influences, recognizing that the interaction between these factors shapes the overall heat island effect.

Emphasizing Physical Elements, Neglecting Functional Activities

Many form indicators of physical elements have been extensively studied in relation to the heat island effect. The related studies focused on two-dimensional landscape pattern indicators and three-dimensional building group indicators. However, there is a notable gap in the literature regarding the mechanisms through which human activity functions influence the heat island effect. While existing research often uses functional differences as mere classification criteria or control variables to investigate the relationship between landscape patterns and the heat island effect, there is a lack of in-depth exploration into how these functional differences in land use contribute to the heat island phenomenon.

Although some scholars have explored the correlation between urban functional elements and the heat island effect from both micro and macro perspectives, there remains a notable gap in explaining how functional use impacts the heat island effect from the essence of human activities. This gap has led to a lack of detailed depiction and mechanism analysis in related research. Urban human activities can be analyzed from at least two dimensions: the type of activity and the intensity of the activity. The type of activity refers to the purpose of people using urban space, encompassing the four major urban functions identified in the "Athens Charter"—residence, work, transportation, and recreation—which are reflected in urban space as distinct land use functions. The intensity of functional use includes two aspects: the spatial extent of a particular function within a given area of the city, which can be represented by metrics such as the floor area ratio, and the level of human activity participation in that space, which can be measured by indicators like employment density, residential population density, and mobile phone signaling density. From the perspective of the formation mechanism of the heat island effect, both dimensions of functional analysis—type and intensity—play a significant role in shaping the heat island effect. The type of activity influences the heat island effect through different land use patterns, while the intensity of activity affects the heat island effect through variations in spatial occupation and human engagement in urban areas. A comprehensive understanding of these factors is essential for advancing research on the heat island effect and developing effective mitigation strategies.

The reason for the relatively few studies in this field is that the impact of function on the thermal environment is mainly mediated by heat emission, which is regarded as a complex process. Personal habits and socio-economic attributes of users can lead to varying amounts of heat dissipation, which complicates the formation of a unified conclusion. Judging from the recent trend in the literature, research on the correlation between function and the heat island effect has gradually shifted from a simplistic focus on land use functions to a more comprehensive analysis that incorporates the type of land use function, the form of building groups, and the landscape pattern. This shift has led to investigations into the relationship between functional intensity and the heat island effect. However, relevant research has yet to reach a conclusive consensus, and there remains controversy over the relationship. Some scholars argue that high-density functional land use exacerbates the heat island effect, while others suggest that low-density residential areas produce more radiation than high-density residential areas and that moderately compact construction is a key strategy for reducing heat island intensity [92].

Emphasizing Correlation Analysis, Downplaying Causal Interpretation

Correlation analysis involves examining the relationship between two or more variables to measure the degree of their association. In research on the heat island effect, scholars frequently quantify various morphological indicators such as land landscape pattern indices, floor area ratios, and green space ratios. While establishing correlation is a necessary step for inferring causality, there is a lack of a systematic rationale for why specific indicators are selected in existing studies. Consequently, much of the research in this field tends to identify correlations without progressing to the explanation of mechanisms or developing a convincing theory of causality. Furthermore, some studies mistakenly equate correlation with causality, attributing the heat island effect solely to observed correlations without deeper causal investigation.

This book posits that a comprehensive understanding of causality must be based on a thorough understanding of correlation. To explore the mechanisms by which urban form influences the intensity of the heat island effect, it is crucial to use existing correlation research as a foundation and combine it with fundamental theories of heat transfer. This approach facilitates the development of a mechanism-based explanation supported by quantitative research.

1.5.6.4 Research Gap of Current Optimizing Strategies

Emphasizing Ontology Derivation, Downplaying Environmental Impact

There are many factors contributing to the formation of the heat island effect, and its causal mechanisms are complex. These factors include both the "internal factors" related to the research object itself and the "external factors" arising from the surrounding environment. Micro-scale numerical simulations often focus on deriving the heat island effect from the inherent attributes of a specific research area, assuming that controlling the urban form of the area can effectively reduce the heat island intensity. However, this approach tends to overlook the impact of external environmental factors. In reality, the strategies for mitigating heat island intensity can vary depending on the external conditions of different surrounding environments.

Emphasizing Convective Heat Transfer, Downplaying Radiant Heat Gain

Past research and practice have suggested that the introduction of suburban cold air is crucial for reducing heat island intensity. By analyzing the overall urban climate environment, this approach identifies "sensitive areas" that significantly affect urban wind corridors, such as interface areas and windward regions near mountains and lakes, and argues that the morphological control of these local key areas is essential for improving the urban thermal environment. These spatial morphological optimization strategies, grounded in the theory of convective heat transfer, aim to reduce heat island intensity from the perspective of the overall urban wind environment. However, in built-up urban areas, high-density construction often prevents the effective influx of external cold air. Consequently, the primary cause of the local heat island effect in such areas is the high-intensity radiant heat capacity of artificial spatial features, which converts significant amounts of solar radiation into heat. Furthermore, variations in heat island intensity within different areas of the built environment are also influenced by differences in spatial morphological elements.

Emphasizing Theoretical Deduction, Lightening Planning Implementation

Most existing studies, after discovering the correlation between spatial morphology and the heat island effect through index values, often propose measures that are disconnected from urban planning management. For instance, common recommendations include statements such as "the floor area ratio is directly proportional to the intensity of the heat island effect, so the floor area ratio should be reduced" or "population density is directly proportional to the heat island effect, so urban population density should be reduced." However, such generalized strategies fail to account for the spatial variation of the effects of spatial morphology on heat island intensity. The impact of a given spatial morphology index on heat island intensity can vary significantly across different urban areas due to spatial differentiation. Thus, a "one-size-fits-all" approach is inadequate for addressing the complexities of urban planning and control. Effective fine-grained planning and management require a more nuanced and scientifically rigorous analysis of spatial morphology and its effects on the heat island effect.

1.5.7 Shortcomings of Existing Research

1.5.7.1 Doubts About the Typicality of the Selection of Urban Form Indicators

Existing studies typically focus on the impact of one or more urban form indicators on the heat island effect [74]. However, given the multitude of available urban form indicators, there is a lack of scientific comparison to determine which indicators have a more prominent effect on heat island intensity. Current research often lacks a rigorous indicator screening process prior to conducting correlation analyses, which undermines the reliability and representativeness of the conclusions drawn. To develop a more robust understanding and provide actionable insights for urban planning, it is essential to perform a systematic evaluation of urban form indicators, identifying those that most significantly influence heat island intensity and ensuring that conclusions are based on thoroughly vetted data.

1.5.7.2 Ignoring the Spatial Instability of Correlation (Spatial Non-stationarity)

Existing studies often assume that the correlation between variables remains spatially stable, implying consistent direction and degree of correlation across different areas within the research scope. However, recent research has shown that this assumption oversimplifies the complexity of the situation. For example, studies have found that the correlation between surface temperature and various environmental indicators exhibits spatial instability [93, 94]. This indicates that the impact of the same urban form indicator on heat island intensity can vary significantly across different regions. Therefore, to more accurately describe the relationship between urban form indicators and heat island intensity, it is essential to conduct a regional analysis.

1.5.7.3 Not Closely Integrated with Urban Planning Management

There is a notable research gap regarding the correlation between the urban heat island effect and the interplay between two-dimensional land use patterns and three-dimensional building forms. Most studies have primarily examined surface cover indicators, such as soil, vegetation, and water bodies, along with remote sensing metrics like NDVI (Normalized Difference Vegetation Index) and NBVI (Normalized Built-Up Index) derived from satellite data [68–70]. While these indicators provide valuable insights into heat island effects, their integration into urban planning management remains challenging. Moreover, research that incorporates statutory control indicators from urban planning frameworks into the analysis is lacking. Existing studies also fall short in translating their findings into practical planning strategies, hindering the development of targeted urban form control methods for different urban regions.

References

1. World Urbanization Prospects 2014 Revision (2014). United Nation, 7
2. The Environmental Protection Agency (EPA) (2014) The heat island effect: heat island impacts. Accessed 27 July 2014
3. Oke TR (2006) The energetic basis of the urban heat island. Q J R Meteorol Soc 455:1–24
4. Steeneveld GJ, Koopmans S, Heusinkveld BG et al (2011) Quantifying urban heat island effects and human comfort for cities of variable size and urban morphology in the Netherlands. J Geophys Res Atmos 116(D20)
5. Guhathakurta S, Gober P (2007) The impact of the Phoenix urban heat island on residential water use. J Am Plann Assoc 73(3):317–329
6. Lundgren K, Kjellstrom T (2013) Sustainability challenges from climate change and air conditioning use in urban areas. Sustainability 5(7):3116–3128
7. Chen Z, Wang H, Ren G (2007) Asymmetric changes in urban heat island intensity in Wuhan. Adv Clim Chang Res 5:282–286
8. Poumadere M, Mays C, Le Mer S et al (2005) The 2003 heat wave in France: dangerous climate change here and now. Risk Anal 25(6):1483–1494
9. Akbari H, Pomerantz M, Taha H (2001) Cool surfaces and shade trees to reduce energy use and improve air quality in urban areas. Sol Energy 70(3):295–310
10. Akbari H, Rosenfeld A, Taha H et al (1996) Mitigation of summer urban heat islands to save electricity and smog. In: 76th Annual meteorological society meeting, Atlanta, GA
11. Liu J, Chen X (2017) Awareness and capability: limits of urban construction. Urban Planning J 1:89–92
12. Chen H, Li B, Zhang W (2015) Correlation study between urban microclimate regulation and block morphological elements. Urban Archit 31:41–43
13. Liang H, Li X, He L (2013) Zoning method for improving urban thermal environment at the city scale in Guangzhou. Urban Planning J 7:107–113
14. Chen J, Zhang Y, Yu Y (2011) Effects of modifiable areal unit problem on spatial autocorrelation. Acta Geogr Sin 66(12):1597–1606
15. Fan C, Myint SW, Kaplan S et al (2017) Understanding the impact of urbanization on surface urban heat islands—a longitudinal analysis of the oasis effect in subtropical desert cities. Remote Sens 9(7):672
16. Li X, Zhou Y, Asrar GR et al (2017) The surface urban heat island response to urban expansion: a panel analysis for the conterminous United States. Sci Total Environ 605:426–435
17. Bonacquisti V, Casale GR, Palmieri S et al (2006) A canopy layer model and its application to Rome. Sci Total Environ 364(1):1–13
18. Djen CS, Jingchun Z, Lin W (1994) Solar radiation and surface temperature in Shanghai city and their relation to urban heat island intensity. Atmos Environ 28(12):2119–2127
19. McPherson EG, Nowak D, Heisler G et al (1997) Quantifying urban forest structure, function, and value: the Chicago urban forest climate project. Urban ecosystems 1(1):49–61
20. Stewart ID, Oke TR (2009) A new classification system for urban climate sites. Bull Am Meteor Soc 90(7):922–923
21. Lo CP, Quattrochi DA (2003) Land-use and land-cover change, urban heat island phenomenon, and health implications. Photogramm Eng Remote Sens 69(9):1053–1063
22. Huang Y (2002) Urban space theory and spatial analysis. Southeast University Press, pp 58–59
23. Conzen MRG (1960) Alnwick, Northumberland: a study in town-plan analysis. Trans Pap (Inst Brit Geogr) 27:iii–122
24. Smailes AE (2007) Geography of towns. Transaction Publishers

25. Oke TR, East C (1971) The urban boundary layer in Montreal. Bound-Layer Meteorol 1(4):411–437
26. Arnfield AJ (2003) Two decades of urban climate research: a review of turbulence, exchanges of energy and water, and the urban heat island. Int J Climatol 23(1):1–26
27. Stewart ID (2011) A systematic review and scientific critique of methodology in modern urban heat island literature. Int J Climatol 31(2):200–217
28. Bai Y, Wang X, Jiang H, Liu S (2013) Research progress on urban heat island effect. J Meteorol Environ 2:101–106
29. Imhoff ML, Zhang P, Wolfe RE et al (2010) Remote sensing of the urban heat island effect across biomes in the continental USA. Remote Sens Environ 114(3):504–513
30. Chen XL, Zhao HM, Li PX et al (2006) Remote sensing image-based analysis of the relationship between urban heat island and land use/cover changes. Remote Sens Environ 104(2):133–146
31. Liu S, Jiang R, Wang C, Wang Y (2014) Observational analysis of spatiotemporal distribution characteristics of urban heat island in Nanjing during summer. J Meteorol Soc 1:19–27
32. Streutker DR (2002) A remote sensing study of the urban heat island of Houston, Texas. Int J Remote Sens 23(13):2595–2608
33. Zhang J, Wang Y (2008) Study of the relationships between the spatial extent of surface urban heat islands and urban characteristic factors based on Landsat ETM+ data. Sensors 8(11):7453–7468
34. Zhang P, Imhoff ML, Wolfe RE et al (2010) Characterizing urban heat islands of global settlements using MODIS and nighttime lights products. Can J Remote Sens 36(3):185–196
35. Aniello C, Morgan K, Busbey A et al (1995) Mapping micro-urban heat islands using landsat TM and a GIS. Comput Geosci 21(8):965–969
36. Kuang W et al (2015) Remote sensing analysis and ecological regulation of urban landmark thermal environment. Science Press, pp 15–16
37. Schwarz N, Manceur AM (2014) Analyzing the influence of urban forms on surface urban heat islands in Europe. J Urban Plann Dev 141(3):A4014003
38. Gallo KP, McNab AL, Karl TR et al (1993) The use of a vegetation index for assessment of the urban heat island effect. Remote Sens 14(11):2223–2230
39. Weng Q (2009) Thermal infrared remote sensing for urban climate and environmental studies: methods, applications, and trends. ISPRS J Photogramm Remote Sens 64(4):335–344
40. Kłysik K, Fortuniak K (1999) Temporal and spatial characteristics of the urban heat island of Łódź, Poland. Atmos Environ 33(24):3885–3895
41. Morris CJG, Simmonds I, Plummer N (2001) Quantification of the influences of wind and cloud on the nocturnal urban heat island of a large city. J Appl Meteorol 40(2):169–182
42. Guo Y, Long B, Liu W, Li M (2006) Observations and preliminary study of Beijing's urban heat island effect. Meteorol Sci Technol 6:656–661
43. Nichol JE, Fung WY, Lam K et al (2009) Urban heat island diagnosis using ASTER satellite images and 'in situ'air temperature. Atmos Res 94(2):276–284
44. Frey CM, Parlow E, Vogt R et al (2011) Flux measurements in Cairo. Part 1: in situ measurements and their applicability for comparison with satellite data. Int J Climatol 31(2):218–231
45. Ben-Dor E, Saaroni H (1997) Airborne video thermal radiometry as a tool for monitoring microscale structures of the urban heat island. Int J Remote Sens 18(14):3039–3053
46. Yu M (2011) Research on comprehensive information mapping of urban thermal environment influenced by remote sensing. Surveying and Mapping Press, Beijing, pp 42–43
47. Voogt JA, Oke TR (2003) Thermal remote sensing of urban climates. Remote Sens Environ 86(3):370–384
48. Gartland L (2011) Heat islands: understanding and mitigating heat in urban areas. Earthscan, Washington, DC
49. Hart MA, Sailor DJ (2009) Quantifying the influence of land-use and surface characteristics on spatial variability in the urban heat island. Theoret Appl Climatol 95(3–4):397–406
50. Taha H (1997) Urban climates and heat islands: albedo, evapotranspiration, and anthropogenic heat. Energy Buildings 25(2):99–103

51. Li W, Bai Y, Chen Q et al (2014) Discrepant impacts of land use and land cover on urban heat islands: a case study of Shanghai, China. Ecol Ind 47:171–178
52. Song YB (2004) Influence of new town development on the urban heat island-the case of the Bundang area. J Environ Sci (China) 17(4):641–645
53. Weng Q, Lu D, Schubring J (2004) Estimation of land surface temperature–vegetation abundance relationship for urban heat island studies. Remote Sens Environ 89(4):467–483
54. Jusuf SK, Wong NH, Hagen E et al (2007) The influence of land use on the urban heat island in Singapore. Habitat Int 31(2):232–242
55. Buyantuyev A, Wu J (2010) Urban heat islands and landscape heterogeneity: linking spatiotemporal variations in surface temperatures to land-cover and socioeconomic patterns. Landscape Ecol 25(1):17–33
56. Frey CM, Rigo G, Parlow E (2007) Urban radiation balance of two coastal cities in a hot and dry environment. Int J Remote Sens 28(12):2695–2712
57. Zheng P, He-ping L, Guo YT et al (2009) Influence of land cover change on land surface temperature in the mountainous city. 地理研究 28(3):673–684
58. Li L, Fu A (1992) On geomorphological causes of urban heat island effect. J Cent China Normal Univ (Nat Sci Ed) 4:115–119
59. Turner BL (2010) Sustainability and forest transitions in the southern Yucatan: the land architecture approach. Land Use Policy 27(2):170–179
60. Weng Q, Liu H, Liang B et al (2008) The spatial variations of urban land surface temperatures: pertinent factors, zoning effect, and seasonal variability. Sel Top Appl Earth Obs Remote Sens IEEE J 1(2):154–166
61. Zhou W, Huang G, Cadenasso ML (2011) Does spatial configuration matter? Understanding the effects of land cover pattern on land surface temperature in urban landscapes. Landsc Urban Plan 102(1):54–63
62. Deng S et al (2014) ENVI remote sensing image processing methods, vol 7. Higher Education Press, Beijing, pp 2–281
63. Rotem-Mindali O, Michael Y, Helman D et al (2015) The role of local land-use on the urban heat island effect of Tel Aviv as assessed from satellite remote sensing. Appl Geogr 56:145–153
64. Chen A, Sun R, Chen L (2012) Applicability of traditional landscape pattern indices in evaluating urban heat island effects. Chin J Appl Ecol 8:2077–2086
65. Cao X, Onishi A, Chen J et al (2010) Quantifying the cool island intensity of urban parks using ASTER and IKONOS data. Landsc Urban Plan 96(4):224–231
66. Xu L, Yue W (2008) A study on thermal environment effect of urban park landscape. Acta Ecol Sinica 28(4):1702–1710
67. Zhang X, Zhong T, Feng X et al (2009) Estimation of the relationship between vegetation patches and urban land surface temperature with remote sensing. Int J Remote Sens 30(8):2105–2118
68. Jenerette GD, Harlan SL, Brazel A et al (2007) Regional relationships between surface temperature, vegetation, and human settlement in a rapidly urbanizing ecosystem. Landscape Ecol 22(3):353–365
69. Ma Y, Kuang Y, Huang N (2010) Coupling urbanization analyses for studying urban thermal environment and its interplay with biophysical parameters based on TM/ETM+ imagery. Int J Appl Earth Obs Geoinf 12(2):110–118
70. Gong A, Chen Y, Li J et al (2007) Relationship between Beijing's urban heat island and land use/cover changes. J Image Graph 12(8):1476–1482
71. Stabler LB, Martin CA, Brazel AJ (2005) Microclimates in a desert city were related to land use and vegetation index. Urban Forest Urban Greening 3(3):137–147
72. Hong S, Lakshmi V, Small EE (2007) Relationship between vegetation biophysical properties and surface temperature using multisensor satellite data. J Clim 20(22):5593–5606
73. Li J, Song C, Cao L et al (2011) Impacts of landscape structure on surface urban heat islands: a case study of Shanghai, China. Remote Sens Environ 115(12):3249–3263
74. Ng E, Yuan C, Chen L et al (2011) Improving the wind environment in high-density cities by understanding urban morphology and surface roughness: a study in Hong Kong. Landsc Urban Plan 101(1):59–74

75. Xu Y, Ren C, Ma P et al (2017) Urban morphology detection and computation for urban climate research. Landscape Urban Plann 167:212–224
76. Li Q, Liu H, Hu F, Hong Z, Li A (2003) Determination of aerodynamic parameters of urban underlying surfaces. Clim Environ Res 4:443–450
77. Ding W, Hu Y, Dou P (2012) Correlation study of urban morphology and microclimate. Archit J 7:16–21
78. Matzarakis A (2005) Country report urban climate research in Germany. IAUC Newsletter 11:4–6
79. Ren C, Ng EY, Katzschner L (2011) Urban climatic map studies: a review. Int J Climatol 31(15):2213–2233
80. Scherer D, Fehrenbach U, Beha HD et al (1999) Improved concepts and methods in analysis and evaluation of the urban climate for optimizing urban planning processes. Atmos Environ 33(24):4185–4193
81. Groß G (2012) Numerical simulation of canopy flows. Springer Science & Business Media
82. Richter CJ, Röckle R (2009) Das numerische simulationsmodell FITNAH
83. Ng E, Kwok K, Sun D et al (2008) Draft urban climatic analysis map. Urban climatic map and standards for wind environment–feasibility study. Working Paper
84. Roth M (2000) Review of atmospheric turbulence over cities. Q J R Meteorol Soc 126(564):941–990
85. Kimura F, Takahashi S (1991) The effects of land-use and anthropogenic heating on the surface temperature in the Tokyo metropolitan area: a numerical experiment. Atmos Environ Part B. Urban Atmos 25(2):155–164
86. Brown MJ, Williams M (1998) An urban canopy parameterization for mesoscale meteorological models. In: Proceedings of the AMS conference on 2nd urban environment symposium, pp 2–7
87. Masson V (2000) A physically-based scheme for the urban energy budget in atmospheric models. Bound-Layer Meteorol 94(3):357–397
88. Li W, Liu H, Zhou X, Qin Y (2003) Analysis of the impact of urban heat island and Taihu Lake on local circulation in the Yangtze River Delta. Sci China (Earth Sci) 2:97–104
89. Liu J, Jiang A, Wang K, Zhu Y, Iguchi R (2006) Development and application of urban microclimate-building coupling evaluation model. J Harbin Inst Technol 1:38–40+55
90. Cheng V (2008) Urban climatic map and standards for wind environment: feasibility study. Methodologies Find User's Wind Comfort Level Survey
91. Stewart ID, Oke TR (2012) Local climate zones for urban temperature studies. Bull Am Meteor Soc 93(12):1879–1900
92. Stone B Jr, Rodgers MO (2001) Urban form and thermal efficiency: how the design of cities influences the urban heat island effect. J Am Plann Assoc 67(2):186–198
93. Li S, Zhao Z, Miaomiao X et al (2010) Investigating spatial non-stationary and scale-dependent relationships between urban surface temperature and environmental factors using geographically weighted regression. Environ Model Softw 25(12):1789–1800
94. Su YF, Foody GM, Cheng KS (2012) Spatial non-stationarity in the relationships between land cover and surface temperature in an urban heat island and its impacts on thermally sensitive populations. Landsc Urban Plan 107(2):172–180

Chapter 2
Urban Heat Island Spatial Pattern and Urban Form Features of Different Heat Island Intensity Zones

This chapter aims to explore the calculation method of heat island intensity and its spatial distribution characteristics, serving as the research foundation for the entire article. It begins by discussing the inversion method of land surface temperature and the calculation method of heat island intensity, as well as the zoning of heat island intensity. Subsequently, it combines the spatial distribution of heat island intensity zoning to summarize the overall spatial pattern characteristics of the heat island from various dimensions, revealing the spatial distribution characteristics of different heat island zones.

2.1 Calculation of Heat Island Intensity

2.1.1 Land Surface Temperature Inversion

The surface temperature inversion in this book employs the atmospheric correction method, also known as the "radiation conduction equation method," using Envi 5.1 software to process Landsat8 TIRS data to derive the surface temperature [1]. The basic principle of this method involves estimating the atmospheric impact on surface thermal radiation, subtracting this atmospheric impact from the total thermal radiation observed by the satellite sensor, and thereby obtaining the surface thermal radiation intensity, which is then converted into the corresponding surface temperature. The processing flow chart is shown in Fig. 2.1.

The thermal infrared radiation brightness value received by the satellite sensor consists of three components: the upward radiation brightness $L\uparrow$, which is the energy of the real radiation brightness of the ground after passing through the atmospheric layer to the satellite sensor; the energy of the downward radiation of the atmosphere reflected after reaching the ground.

The steps for surface temperature inversion are as follows:

Y. Lu, *Urban Form and Urban Heat Island*, Urban Sustainability,
https://doi.org/10.1007/978-981-96-3541-2_2

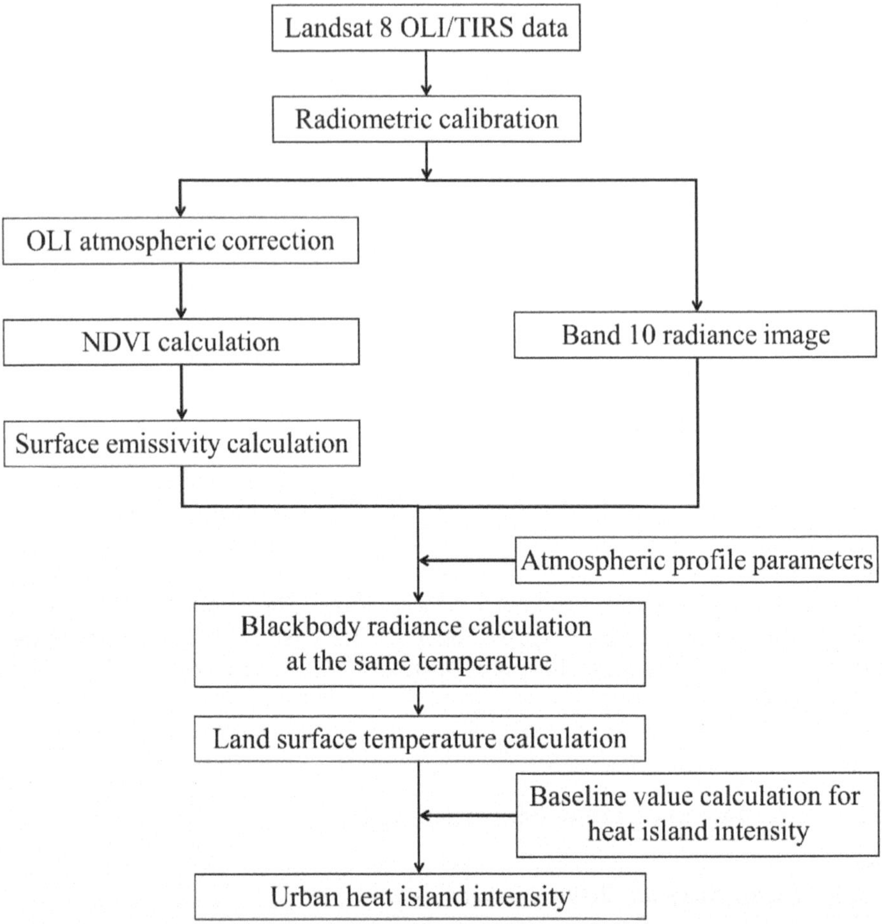

Fig. 2.1 Land surface temperature calculation flow chart based on atmospheric correction method

2.1.1.1 Image Radiometric Calibration

Radiometric calibration involves converting the voltage or digital quantization recorded by the satellite sensor into absolute radiance values or relative values related to physical quantities such as ground (apparent) reflectance or surface (apparent) temperature. The calibration tool in ENVI software (Radiometric Calibration) is used to calibrate the image data into radiance values. The calibration is performed using the following formula:

$$L = Gain \times DN + offset$$

where Gain and Offset are the inherent parameters for each band.

2.1.1.2 Calculation of Surface Emissivity

The mixed model method is used to calculate surface emissivity, utilizing the NDVI threshold method proposed by Sobrino et al. [2]. This approach assumes that the surface is composed of vegetation and bare ground, with NDVI used for surface classification. The process involves the following steps:

(1) When NDVI < 0.2, it is considered to be completely covered by bare ground, and the surface emissivity is taken as the typical emissivity value of bare ground, 0.973.
(2) When 0.2 ≤ NDVI ≤ 0.5, it is considered that the pixel is a mixed pixel composed of vegetation and bare ground, and the surface emissivity is calculated by the simplified formula:

$$\varepsilon = 0.004 \times P_v + 0.986$$

(3) When NDVI > 0.5, it is considered to be completely covered by vegetation, and the surface emissivity is taken as the typical emissivity of vegetation, 0.986.

Where, Pv is the vegetation coverage, calculated by the following formula:

$$P_v = \frac{NDVI - NDVI_{Soil}}{NDVI_{veg} - NDVI_{Soil}}$$

where, NDVI is the normalized vegetation index, NDVI Soil is the NDVI value of completely bare soil or non-vegetation covered areas, and NDVI Veg represents the NDVI value of pixels completely covered by vegetation, i.e., the NDVI value of pure vegetation pixels. Here, the simplified vegetation coverage calculation model in ENVI is used to calculate the NDVI value, and then the vegetation coverage image is obtained through band calculation.

$$\text{NDVI} = \frac{\rho \times NIR - \rho \times RED}{\rho \times NIR + \rho \times RED}$$

$$(\text{b1 gt } 0.7) \times 1 + (\text{b1 It } 0.05) \times 0 + (\text{b1 ge } 0.05 \text{ and b1 le } 0.7) \times \frac{b1 - 0.05}{0.7 - 0.05}$$

where, b1: NDVI.

Then substitute the vegetation coverage image to get the surface emissivity image.

$$0.004 \times \text{b1} + 0.986$$

where, b1: vegetation coverage image.

2.1.1.3 Calculation of Blackbody Radiance and Surface Temperature

Query the NASA website (http://atmcorr.gsfc.nasa.gov), enter the imaging time: 2014-10-06, 02:55 (Greenwich Mean Time) and the central latitude and longitude (Lat: 30.306, Lon: 113.832), as well as other corresponding parameters, to get the atmospheric profile information as follows:

1. The transmittance τ of the atmosphere in the thermal infrared band: 0.83
2. Upward radiance $L\uparrow$ of the atmosphere: 1.35 $W/(m^2 \cdot sr \cdot \mu m)$
3. Downward radiance $L\downarrow$ of the atmosphere: 2.25 $W/(m^2 \cdot sr \cdot \mu m)$

According to the following formula, the blackbody radiation brightness image at the same temperature is calculated.

$$\frac{b2 - 1.35 - 0.83 \times (1 - b1) \times 2.25}{0.83 \times b1}$$

where

b1 Surface emissivity image
b2 Band10 radiance image

Based on the following formula, the surface temperature image in degrees Celsius is calculated.

$$\frac{1321.08}{a \log \frac{774.89}{b1} + 1} - 273$$

where, b1 is the blackbody radiation brightness image at the same temperature, and the surface temperature image is subsequently obtained.

Although the selected remote sensing images exhibit minimal cloud cover, some areas in the study region are still affected by clouds, leading to distortions in the surface temperature inversion. To minimize these errors, areas with surface temperatures below 26 °C, which are likely influenced by cloud cover, are excluded from the study area. This approach ensures more accurate temperature inversion and reliable analysis of the heat island effect.

2.1.2 Calculation of Urban Heat Island Baseline and Zoning According to Urban Heat Island Intensity

The calculation of urban heat island intensity uses the calculation method of the Urban Heat Island Intensity Index (UHII) [3], that is:

$$UHII_i = T_i - \frac{1}{n}\sum_{1}^{n} T_{crop}$$

where $UHII_i$ represents the heat island intensity for the i-th pixel in the image, Ti is the surface temperature of that pixel, and n is the total number of valid pixels in the suburban farmland. Tcrop, the surface temperature in the suburban farmland, is used as the baseline for heat island intensity in this book. This baseline temperature, obtained as the average surface temperature of all agricultural and forestry land (E2) on the urban fringe, is set at 26.98 °C. Any area with a surface temperature exceeding 26.98 °C is identified as having a heat island effect, referred to in this book as a "heat island."

Using this baseline temperature and the natural breaks method, the study classifies the heat island intensity into five categories based on observed surface temperatures: extremely strong heat island area, strong heat island area, relatively strong heat island area, weak heat island area, weak cold island area and strong cold island area (Table 2.1 and Fig. 2.2).

Table 2.1 The main context of the existing literature regarding to urban form

Zoning name	Suface temperature (°C)	Heat Island intensity (°C)
Strong cold island area	Less than 26.10	–
Weak cold island area	26.11–26.98	–
Weak hot island area	26.98–28.30	0–1.34
Strong hot island area	28.30–29.61	1.34–2.63
Strong hot island area	29.61–30.93	2.63–3.95
Strong hot island area	Greater than 30.93	More than 3.95

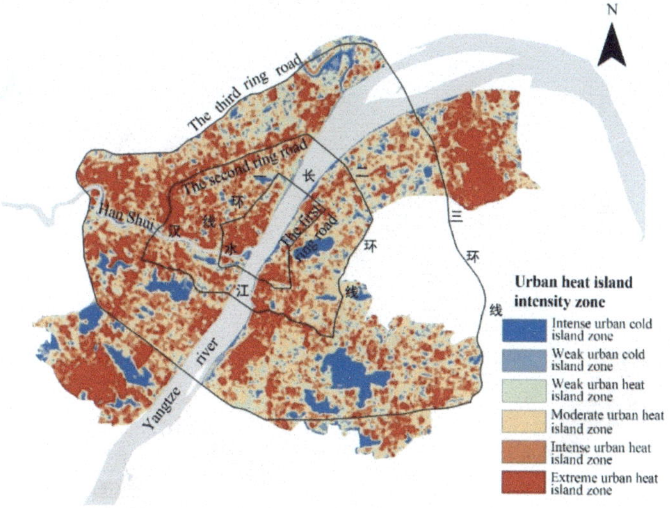

Fig. 2.2 Spatial layout of different urban heat island zones. Different colors indicate different urban heat island zones

2.2 Overall Spatial Pattern Characteristics of Heat Islands

2.2.1 "U" Shaped Ring Layer Differentiation with "High Inner and Outer, Middle Collapse"

Using the "Feature to Point" function in ArcGIS software, the geometric center of the study area is determined as the reference point for ring layer analysis. A ring layer distribution map is then created with a distance interval of 500 m, ranging from 0 to 20,000 m from the center (Fig. 2.3). This map intersects with the study area to calculate the average surface temperature and heat island intensity for each of the 41 ring layers.

The analysis reveals that as the distance from the geometric center increases (Fig. 2.4), the heat island intensity undergoes four distinct stages: Core Ring (0–3500 m), Inner Ring (3500–9500 m), Middle Ring (9500–16,000 m), and Outer Ring (16,000–20,000 m).

2.2.1.1 Core Ring—"High and Low Fluctuation" Stage

Within the core ring, the heat island intensity exhibits notable fluctuations. It peaks at 2.54 °C at the 500-m ring layer, then decreases to a low of 2.00 °C at the 2000 m point before rising again to 2.65 °C at the 3500-m ring layer.

From a micro-spatial perspective, most of the built-up areas are characterized by extremely strong and strong heat island effects. However, the Yangtze River

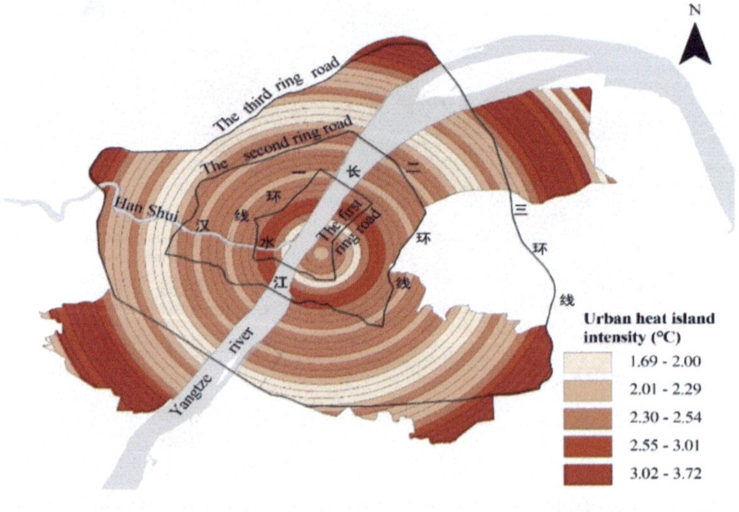

Fig. 2.3 Urban heat island variation on urban–rural gradient

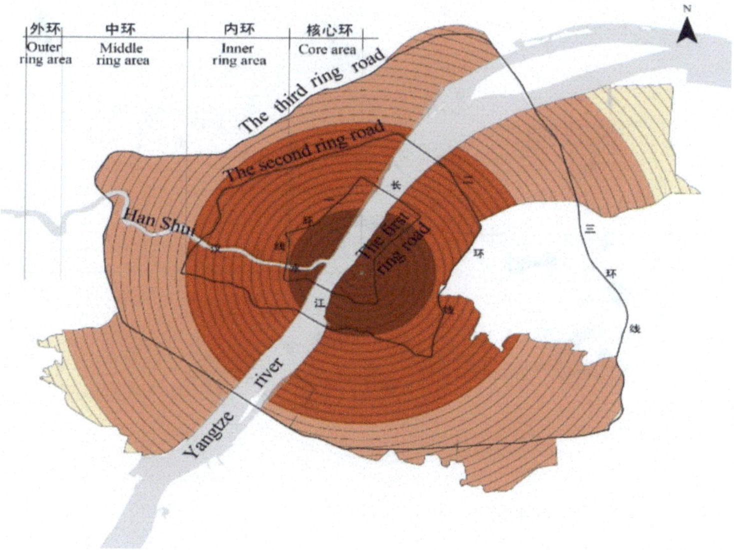

Fig. 2.4 Spatial layout of core area, inner ring area, middle ring area and outer ring area

significantly mitigates the heat island effect in the riverside areas of Wuchang and Hanyang. As one moves further from the riverbank, the surface temperature gradually increases. In contrast, the riverside area of Hankou does not exhibit a similar heat island intensity buffer zone. High-intensity heat island areas are concentrated near the Yangtze River shoreline, with a minimal buffer zone around Sand Lake, which is also surrounded by a strong heat island zone (Fig. 2.5).

This region encompasses the core functions of Wuhan city, including major city centers such as Hongshan Square and Jianghan Road (Fig. 2.5). The built-up areas here are dense and compact, creating a localized high-intensity heat island effect. Despite this, the presence of the Yangtze River, Han River, and Sand Lake within the core ring introduces an open and closed ecological pattern that reduces the average heat island intensity for some ring layers and limits the further expansion of high-intensity heat islands.

2.2.1.2 Inner Ring—"Continuous Stability" Stage

The heat island intensity in the inner ring, which extends from 4000 to 8000 m from the geometric center of the main city, remains relatively stable at approximately 2.40 °C (Fig. 3.4). Within this inner ring, the reduction in heat island intensity is slightly less pronounced in the Hankou and Hanyang areas compared to Wuchang.

In Hankou and Hanyang, the extremely strong and strong heat island areas are mostly contiguous, with the most intense heat island effects occurring along the

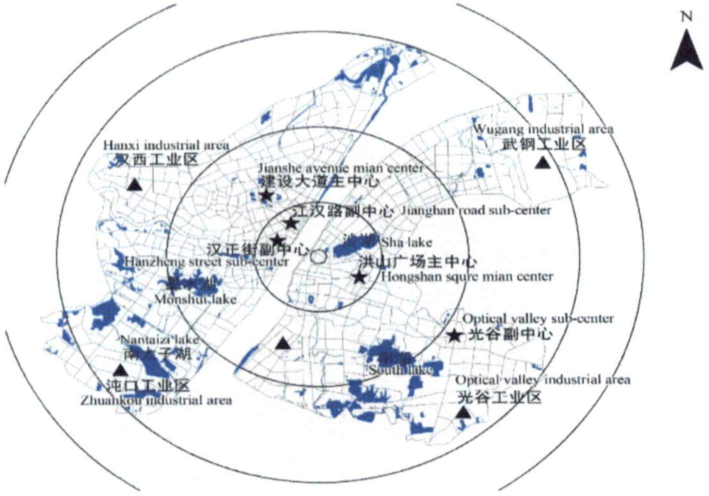

Fig. 2.5 Main water bodies and public service centers

Second Ring Road and both banks of the Han River. Weak heat island and cold island areas are found primarily in scattered patches within the Second Ring Road. Conversely, in Wuchang, the contiguous area of extremely strong heat islands is confined mainly to the southern riverside region. Here, large water bodies like Nanhu and the northern riverside significantly lower heat island intensity, while other areas exhibit considerable fluctuations in heat island intensity over small distances. Overall, the extent of the extremely strong heat island effect in Wuchang is relatively limited.

The analysis reveals that within the inner ring, heat island intensity does not show a significant gradient of decrease from the center to the periphery. This lack of a gradient is due to the presence of strong cold island effects at Nanhu and Moshui Lake, which are located on the outer edge of the inner ring and contribute to the decrease in heat island intensity in the 8500–9500 m range of the ring layers.

2.2.1.3 Middle Ring—"Rapid Ascending" Stage

In the middle ring, which extends from 9500 to 16,000 m and lies almost entirely outside the Second Ring Road, heat island intensity exhibits significant periodic fluctuations but overall shows a marked increase from the inner to outer edges. Specifically, the intensity rises from 1.89 °C at 9500 m to 3.71 °C at 16,000 m, reaching the maximum heat island intensity observed across all ring layers. The average increase in heat island intensity across the middle ring is 0.09 °C per ring layer.

Within this ring, Nanhu and Nantaizi Lake, situated in the 9500–10,500 m range, provide a cooling effect that reduces surface temperatures. Despite this localized

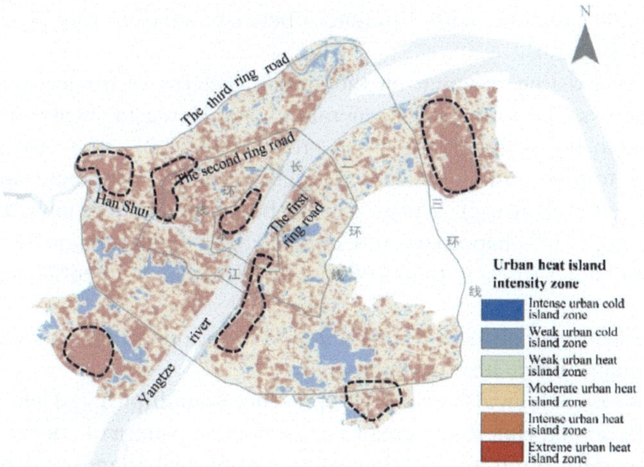

Fig. 2.6 Spatial layout of hot area

cooling, the majority of the middle ring remains a strong heat island area. Specifically, the south bank of Nanhu features low-rise educational and research facilities, creating a more dispersed patchy heat island pattern dominated by relatively strong and weak heat island areas. Conversely, Nantaizi Lake, surrounded by extensive low-rise industrial land, forms a dense, extremely strong heat island area near the shoreline.

Adjacent to the Third Ring Road, the concentrated industrial zones of Wugang and Zhuankou contribute to a very high heat island intensity, which peaks in the 14,000–16,000 m range (Fig. 2.6). Additionally, the Guanggu South area in the southeast part of the study region also exhibits a significant strong heat island presence, though it is interspersed with numerous cold islands, reflecting a more complex landscape pattern.

In summary, the heat island intensity increases significantly from the inner to outer regions. This increase is influenced by the localized cooling effects of Nanhu and Nantaizi Lake and the dense industrial development around the Third Ring Road. The combination of these factors leads to the highest observed heat island intensities at the outer edge of the middle ring.

2.2.1.4 Outer Ring—"Large Fluctuation" Stage

This area, located on the edge of the main city, exhibits a smaller study range for each ring layer and a limited number of ring layers. Furthermore, there are significant differences in land use functions across these ring layers: industrial land dominates the inner ring layers, while green spaces become more prevalent in the outer layers. As a result, the heat island intensity in each ring layer shows greater fluctuations compared to other regions, with differences between adjacent ring layers reaching up to 1.52 °C (Fig. 3.4).

A review of existing research on the spatial patterns of urban heat islands in major cities across China revealed a general trend of decreasing heat island intensity from the core to the periphery of urban areas (Table 2.2). Typically, this spatial differentiation follows a monotonically decreasing pattern with increasing distance from the city center. However, Wuhan's urban heat island pattern deviates from this norm, presenting a "U"-shaped structure characterized by high-intensity heat islands in the inner and outer rings, with a "collapse" or decrease in intensity in the middle layers.

Two main factors contribute to this unique "inner, outer high, middle collapse" pattern in Wuhan's urban heat island intensity:

The first one is ecological Barriers. The presence of multiple large lakes, including the Yangtze River and Han River, creates an ecological pattern of "big rivers and big lakes." This pattern limits the continuous spread of heat islands and prevents the accumulation of artificial heat. Consequently, while the core and inner rings have the necessary conditions for forming large-scale, high-intensity heat islands, the heat island intensity decays in the middle ring, resulting in a "valley" or "low point" in the heat island intensity distribution between 9500 and 10,500 m.

The second one is industrial Heat Emission. Large-scale industrial areas on the edge of the main urban zone, such as those in Wugang and Zhuankou, not only generate intense heat island effects themselves but also influence the surrounding non-industrial areas. This industrial heat emission causes a significant increase in heat island intensity in the middle ring, peaking at 3.71 °C at 16,000 m and establishing the highest heat island intensity within the study area (Fig. 3.8).

These factors collectively contribute to Wuhan's distinctive heat island intensity pattern, contrasting with the more uniform land surface temperature decrease in the urban–rural direction observed in other cities.

2.2.2 *Spatial Pattern of Hot Area*

From the perspective of heat island intensity zoning, some extremely strong heat island areas and strong heat island areas are closely related to urban roads, including Hanyang Avenue on the south bank of the Han River and Jiefang Avenue on the north bank, the northwest section of the Second Ring Road, Longyang Avenue in Hanyang north–south, and Baishazhou Avenue near the south of Wuchang along the Yangtze River (Fig. 2.7). The heat island intensity of the regulatory planning units near these city roads is higher, and it is clearly distinguished from the regulatory planning units far from these roads.

Table 2.2 Comparative analysis of spatial pattern characteristics of heat islands in big cities in China

Case city	Space types on which the analysis is based	Time node/segment	Spatial pattern characteristics of heat island	Relationship between the spatial pattern of heat island and the urban spatial elements
Beijing	Specific region circle layer (divided by concentric circles of equal area)	2012	The single center circle structure is prominent, and the heat island strength gradually decreases from the core area to the periphery	When the proportion of construction land area exceeds 50%, this area will produce significant heat island phenomenon; the footprint of urban heat island area is basically consistent with the scope of construction land
Nanjing	Specific area circle layer (divided by urban area and suburbs)	2010	The urban high temperature area is vast, forming a closed heat island center around the city; the heat island strength slowly decreases from the center to the periphery	The high temperature center is located in the densely urbanized area, while the east direction of the city becomes the low temperature center due to the influence of the Xuanwu Lake water body and the Purple Mountain woodland
Hangzhou	Specific urban functional areas (divided by business district, area along railway and waterfront area)	2007	The heat island in the core area is the strongest, radiating outward and gradually weakening; there is one cold island on each side of the main city, respectively scenic area and reserved land	Centralized commercial areas, industrial and storage concentration areas, and the areas along the railway have strong heat islands, and water bodies, green space, mountains and farmland form low-temperature areas

(continued)

Table 2.2 (continued)

Case city	Space types on which the analysis is based	Time node/segment	Spatial pattern characteristics of heat island	Relationship between the spatial pattern of heat island and the urban spatial elements
Shanghai	Space lattice	2002	From the inside of the central city to the periphery, the surface temperature gradually presents a layer expansion pattern of "high and low values", and shows an asymmetric distribution pattern in the east–west upward	Industry, land cover type, residential area type and heat island intensity were related significantly
Xian	Specific urban functional areas (divided by street level)	2008	There is a strong heat island center and a weak heat island center, and the intensity of heat island effect is decreasing along the city center; it is obvious horseshoe distribution, strong in the middle and weak around it	The relationship between leading urban function, green space underlying surface, water resources, building density, industrial and commercial agglomeration degree and heat island intensity is significant

2.2.3 *Aggregation of Hot and Cold Urban Areas*

According to the First Law of Geography, geographical phenomena are more closely related to nearby phenomena than to those farther away, a concept known as spatial autocorrelation. Similarly, the heat island effect, a widespread geographical phenomenon, exhibits spatial connectivity where the thermal environments of adjacent areas influence each other through convective heat transfer. To analyze this spatial autocorrelation, the Global Moran's I tool available in ArcGIS is employed. This tool evaluates spatial patterns by constructing a 500 × 500 m spatial grid and analyzing the relationship between the positions of these spatial units and their corresponding average surface temperatures. The Global Moran's I tool determines whether the observed pattern is a clustering pattern, a dispersion pattern, or a random

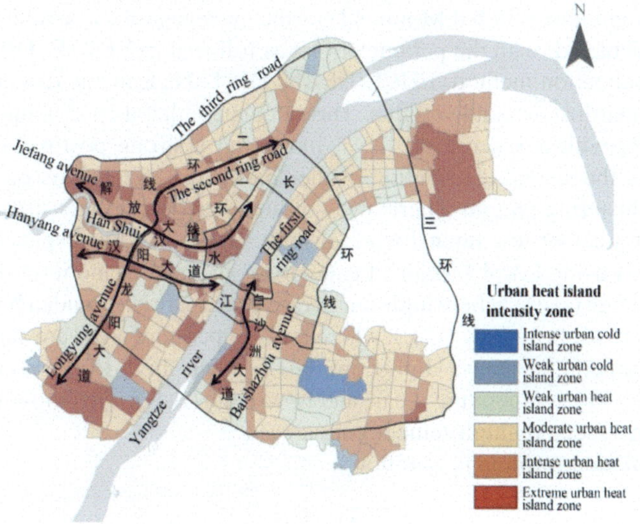

Fig. 2.7 Spatial out of extreme hot area and intensive hot area

pattern. By assessing these patterns, the tool provides insights into the spatial distribution of heat island intensity and identifies whether areas of high or low intensity are grouped together or spread out across the study area.

The calculation formula is as follows:

$$I = \sum_{i=1}^{n}\sum_{j=1}^{n}\frac{x_i - \overline{x}}{s^2}\sum_{i=1}^{n}\sum_{j=1}^{n}W_{ij}$$

In the formula: x_i is the average surface temperature observation value of spatial grid i; W_{ij} is the spatial weight matrix, where spatial neighbors are 1, and non-neighbors are 0.

The formula for the Z test value is:

$$Z(I) = \frac{I - E(I)}{\sqrt{Var(I)}}$$

where, E(I) is the mathematical expectation under the hypothesis of spatial non-agglomeration; Var(I) is the variance. When Z(I) value is positive and significant, it indicates that there is a positive spatial autocorrelation in the region, that is, high-high clustering or low-low clustering. When Z(I) value is negative and significant, there is a negative spatial autocorrelation, that is, high-low clustering or low–high clustering. When Z(I) is 0, the observed values are distributed independently and randomly.

For the calculation of Moran's I Index, the full-domain 500 × 500-m spatial grid of the management unit is used to reflect all spatial grids and capture the spatial autocorrelation characteristics, especially around cold island areas. The global spatial autocorrelation index (Global Moran's I) of the average surface temperature observations is calculated, with the parameters set as follows: INVERSE_DISTANCE as the conceptualization method, EUCLIDEAN_DISTANCE as the distance method, and non-standardized spatial weights. The calculated global spatial autocorrelation index is 0.66, with a Z score of 43.97, indicating a strong positive spatial autocorrelation in the spatial distribution of the heat island effect. This result shows a significant clustering pattern where areas of high heat island intensity are grouped together and areas of low intensity are also clustered. The local spatial autocorrelation analysis using Local Moran's I categorizes the heat island intensity into four types: High-High (H-H), where high-temperature areas are surrounded by other high-temperature areas; Low-Low (L-L), where low-temperature areas are surrounded by other low-temperature areas; High-Low (H-L), where high-temperature areas are surrounded by low-temperature areas; and Low-High (L-H), where low-temperature areas are surrounded by high-temperature areas.

The calculation formula is as follows:

$$I_i = \frac{(x_i - \overline{x})}{\sum_i (x_i - \overline{x})} \sum_j W_{ij}\left(x_j - \overline{x}\right)$$

In the formula, x_i and x_j represent the average surface temperature of spatial grid i and j, respectively, and w_{ij} is the corresponding element of the spatial weight matrix w, with a total of n spatial grids in the study area. The value of I_i ranges from [−1, 1], where the sign indicates the type of spatial autocorrelation, the absolute value indicates the strength of autocorrelation, and 0 represents no spatial autocorrelation.

According to calculations, the "High-High" and "Low-Low" spatial correlation types in the study area each account for 9.09% and 8.67%, respectively, indicating that the heat island effect exhibits a significant positive correlation in local areas. At the same time, high-temperature and low-temperature areas are relatively isolated, which explains the absence of "High-Low" and "Low-High" types. The spatial grids associated with the "High-High" type primarily include three types of functional spaces: ① Concentrated and large-area industrial areas, including Wugang, Duankou, and two major industrial zones, as well as the Baishazhou industrial Area; ② High-density commercial and residential mixed areas, such as the Han Zheng Street in Hankou district; ③ Concentrated wholesale market areas, such as the Hanxi Building Materials Market. Among these, concentrated industrial areas occupy the vast majority. The "Low-Low" spatial types are relatively singular, mostly consisting of large-scale water bodies or areas surrounding these water bodies, with the largest "Low-Low" concentration area located in Nanhu Lake (Fig. 2.8).

Although the overall area of the heat island is relatively large and densely distributed, non-heat island areas dominated by large-scale water bodies and mountains have created a significant isolation effect on the heat island areas, resulting in relatively concentrated cold island regions. After excluding the grids with a surface temperature lower than 26.98 °C, the heat island area covers most of the study area, accounting for 87.56%. Lakes such as Nanhu Lake, Shahu Lake, South Taizi Lake,

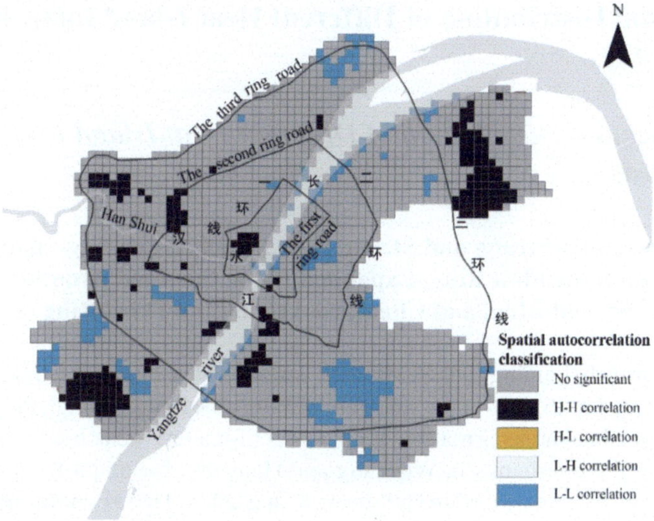

Fig. 2.8 Local spatial autocorrelation of urban heat island intensity

and Moshui Lake do not exhibit a heat island effect. Some built-up areas around them also lack a heat island effect. Conversely, cold and hot island areas formed by Flat Mountain, Turtle Mountain, Hongshan, Luojia Mountain, and Ma'anshan are arranged from west to east, traversing the main urban area of Wuhan (Fig. 2.9).

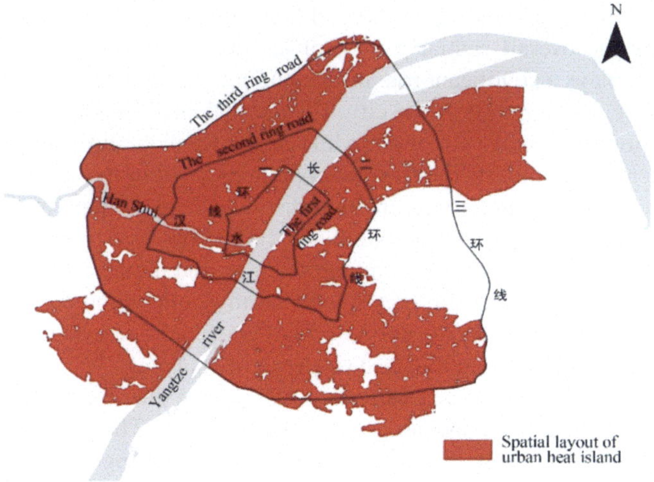

Fig. 2.9 Spatial layout of urban heat island

2.3 Spatial Distribution of Different Heat Island Intensity Areas

2.3.1 Location Distribution of Different Heat Island Intensity Areas

2.3.1.1 Extremely Strong and Strong Heat Island Areas: Aggregated Within the First Ring, Expanded Along the Road Around the Second Ring, and Clustered Around the Third Ring

The spatial grids belonging to the extremely strong and strong heat island areas total 980, accounting for 46.12% of the study area. Within the first ring road, the proportion of strong heat island areas is notably high. Except for a few weak heat island regions located along the riverbanks in Wuchang and Hanyang, the majority of these areas are occupied by extremely strong and strong heat islands. These conditions are highly concentrated in the historical urban centers of Hankou and Wuchang, demonstrating a pronounced spatial autocorrelation.

As we move outward from the first ring road, the spatial distribution of heat island intensities rises. Around the second ring road, the clustering of extremely strong and strong heat island areas diminishes, transitioning from a dense, block-like distribution to a more linear arrangement along major roads. In the vicinity of the third ring road, strong heat island areas exhibit distinct, plate-like patterns, particularly in the Wugang and Zhuankou industrial districts. These areas appear relatively isolated and disconnected from the extremely strong and strong heat island regions, giving the impression of "floating" on the periphery of the main city.

Examining the overall spatial pattern of heat island intensity reveals that the Yangtze River plays a significant role in separating the heat islands in the Wuchang area from those in other parts of the city. The thermal environment in the Wuchang district functions in relative isolation, while in the Hankou and Hanyang regions, there is a clear continuity of the thermal conditions. Within the Hankou-Hanyang area, the strong heat island zones create a distinctive spatial pattern described as "two horizontals, one vertical, and one block". The "two horizontals" refer to the strong heat island axes extending along the southern and northern banks of the Han River, while the "one vertical" represents the north–south alignment of heat island areas across the Hankou-Hanyang corridor. The "one block" describes the Zhuankou industrial area located beyond the third ring road of Hanyang.

In contrast, the Wuchang area features a "T"-shaped distribution of strong heat island regions alongside a concentrated heat island block. The "T"-shaped pattern consists of the riverside zones in the southern part of Wuchang and the areas along the western end of Xiongchu Avenue, while the heat island block is centered around the Wugang industrial area. This configuration highlights how the spatial distribution of heat islands varies significantly between different urban sectors, shaped by both natural and anthropogenic factors.

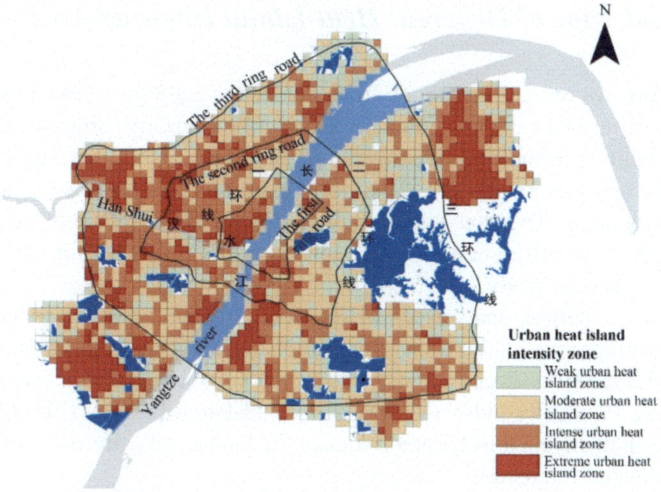

Fig. 2.10 Spatial layout of different urban heat island intensity zones in fishnets

2.3.1.2 Strong and Weak Heat Island Areas: Riverside and Around the Lake, Located on the Edge of the Heat Island

The spatial grids belonging to the relatively strong and weak heat island areas account for 17.25% and 31.87% of the study area, respectively. Compared to the extremely strong and strong heat island areas, the relatively strong heat island areas are more dispersed and exhibit a spatial distribution that is closely associated with the structure of the strong heat island areas. These relatively strong heat island areas are typically located around the peripheries of the strong heat island zones, often filling the "gaps" between adjacent but non-contiguous strong heat island regions (Fig. 2.10).

In contrast, the weak heat island areas display a more concentrated distribution. They are primarily found along riverbanks and around large water bodies such as Nanhu Lake, Shahu Lake, Moshui Lake, Donghu Lake, and Shenshui Lake. The weak heat island areas are also notably present in the "gaps" between different relatively strong heat island regions, further illustrating the spatial interconnections of these areas.

This spatial arrangement of heat island intensities reveals distinct patterns of heat distribution across the study area. The relatively strong heat island areas tend to be more scattered and linked to the strong heat island areas, while the weak heat island areas are more concentrated around natural water bodies and in the spaces between stronger heat island zones. This pattern emphasizes the role of large-scale environmental features, such as rivers and lakes, in moderating heat island effects and creating spatial variability in heat island intensity.

2.3.2 *Landscape of Different Heat Island Intensity Area*

This book borrows research methods from landscape ecology to analyze the patch shape within different heat island intensity areas in Wuhan City using surface temperature grids obtained from Landsat 8 inversion (pixel size: 30 × 30 m). By calculating various landscape pattern indices, This book explores the spatial structure of heat island intensity areas. The analysis is conducted using the software of Fragstats 4.2, focusing on the four different heat island intensity categories within the heat island areas as the objects of study.

The landscape pattern analysis for different heat island intensity areas examines their patch number and degree of agglomeration. The selected indices include Type Proportion (CPLAND), Largest Patch Index (LPI), Average Patch Area (AREA_MN), Average Proximity Index (PROX_MN), and Patch Density (PD) (Table 2.3). These indices help reveal the characteristics of the landscape patterns associated with various heat island intensities.

To uncover the overall layout characteristics of the landscape patterns, the indices for each heat island intensity area are computed across four major ring layers: the core ring, inner ring, middle ring, and outer ring. The statistics reveal that both the Strong Heat Island Areas and Relatively Strong Heat Island Areas exhibit not only a dominant proportion of the total area but also larger individual patch sizes, a pattern particularly pronounced in the inner and middle ring layers. In contrast, the Weak Heat Island Areas and Extremely Strong Heat Island Areas are characterized by smaller individual patches.

It is observed that the Strong and Relatively Strong Heat Island Areas display a higher degree of patch proximity, indicating a tendency for these patches to be clustered together. Conversely, the Weak and Extremely Strong Heat Island Areas

Table 2.3 Main landscape metrics

Name of index	Indicator significance	Unit
Type ratio (CPLAND)	Area of the patch type and the ratio of area occupied by each type	%
Maximum patch index (LPI)	Proportion of maximum plaques to the study area	%
Mean patch area (Area_Mn)	Area average of the patch	ha
Average proximity index (PROX_MN)	Distance proximity of similar patches within a 500 m search radius	–
Shape index (Shape_Am)	The complexity of the patch shape has a value of 1 as the patch shape approaches the square, and increases as the patch shape becomes more complex	–
Patch density (PD)	The degree of fragmentation of each type of patch	–

across all ring layers exhibit more dispersed individual patch characteristics, with loose connections among similar patches and less continuity in the landscape pattern.

Overall, this book demonstrates that different heat island intensities are associated with distinct landscape patterns. The Strong and Relatively Strong Heat Island Areas are marked by larger, more clustered patches, whereas the Weak and Extremely Strong Heat Island Areas feature smaller, more dispersed patches. This analysis provides insights into the spatial pattern differentiation of heat islands and highlights the influence of landscape structure on the distribution of heat island intensity in Wuhan City.

2.3.3 Patch Association Between Different Heat Island Intensity Areas

The spatial relationship among different intensity heat island areas exhibits a spatial nest pattern. In this arrangement, the Extremely Strong Heat Island Areas serve as the "core" of the heat island phenomenon, forming well-defined axes and blocks that establish the spatial context for the other heat island intensity areas. These areas are highly concentrated and recognizable. Surrounding these core areas, the other heat island areas are organized in a layered manner, with the spatial centers of less intense heat island areas often situated at the peripheries of the more intense ones. This configuration creates a tightly interlinked structure where each heat island intensity area functions as a buffer zone for the preceding intensity level, forming a nested and gradient buffer structure between different heat island intensities.

The spatial distribution begins with the "Extremely Strong Heat Island Area" at the highest intensity level, with subsequent heat island intensity areas expanding outward along the edges of the previous levels, gradually transitioning towards the Cold Island Areas. This "gradual nesting" ensures that there is generally no "leapfrogging" of heat island intensity within relatively close spatial distances; rather, there is a smooth transition from high to low intensity with a pronounced buffering effect. To objectively verify this "nested structure" and "gradient buffering" phenomenon, we create 1000-m buffer zones around the outer edges of each heat island intensity area and intersect these zones with average surface temperature data based on spatial grids using ArcGIS. The results show that the buffer zones primarily consist of adjacent heat island intensity areas, with minimal cross-level intensity interactions.

Additionally, although the "Strong Cold Island Areas" and "Weak Cold Island Areas," dominated by large-scale lakes, are not considered part of the heat island area, they significantly influence the overall spatial pattern and intensity differentiation of the heat island effect. From a macro perspective, excluding the large-scale lakes, the study area predominantly features heat island effects. These lakes effectively limit the spatial extent of the heat island areas. Furthermore, the presence of these lakes contributes to a gradual transition between the Strong Heat Island Areas and the Cold Island Areas, thereby "softening" the edges of the heat island regions. This transition reduces the overall heat island intensity and plays a crucial role in mitigating the heat island effect.

2.4 Summary

This chapter addresses two primary questions: (1) What are the characteristics of the spatial pattern of heat islands in Wuhan? and (2) What are the spatial distribution rules and landscape patterns of different heat island intensity areas?

Firstly, the analysis reveals that heat islands are widely distributed across the main urban area, with the spatial differentiation of heat island intensity exhibiting several distinct characteristics. These include a "double high inside and outside, middle collapse U-shaped" ring differentiation, multi-axis extension, and local high-temperature aggregation, low-temperature aggregation. Specifically, 87.56% of the study area experiences a heat island effect, excluding only large lakes and mountains. The spatial pattern of heat island intensity is explored through three key perspectives: the ring pattern, the axis pattern, and the spatial aggregation-dispersion pattern.

1. Buffer Ring Pattern of Heat Island Intensity: The heat island intensity in Wuhan's main urban area demonstrates a four-stage ring differentiation from the core to the periphery: the core ring (0–3500 m) shows fluctuations, the inner ring (4000–9000 m) displays continuous stability, the middle ring (9500–16,000 m) experiences rapid increases, and the outer ring (16,500–20,000 m) exhibits significant fluctuations. This pattern presents a "double high inside and outside" spatial distribution rule, which contrasts with the common "monotonic decrease from the center to the periphery" observed in other major cities. The unique spatial distribution in Wuhan could be attributed to the large number of lakes in the central urban area, which have a cooling effect on the surrounding built-up areas. Also, the industrial areas on the periphery of urban areas intensify the heat island effect in the outer ring layers.
2. Axis Pattern of Heat Island Intensity: The analysis of main city roads extending from the core area within the first ring to the peripheral ring layers appears to be a relatively concentrated high-temperature region. This distribution forms an axial spatial pattern, where high-temperature areas are aligned along specific routes radiating from the core to the periphery.
3. Spatial Aggregation-Dispersion Pattern: Using Moran's I index for global spatial autocorrelation analysis, it is found that the spatial distribution of heat island intensity exhibits strong positive spatial autocorrelation. This indicates that extremely strong and strong heat island areas tend to cluster together, creating distinct "high-high" aggregation patterns, while high-temperature and low-temperature areas exhibit tendencies to aggregate in contiguous blocks.

Secondly, the study identifies distinct locational characteristics, patch morphology, and spatial associations for different heat island intensity areas:

1. Locational Distribution Pattern: The extremely strong and strong heat island areas are concentrated within the first ring, expand along the roads around the second ring, and form plate-like distributions around the third ring. In contrast, the relatively strong and weak heat island areas are prominently located along rivers and lakes, typically situated at the edges of the built-up areas.
2. Patch Morphology Pattern: The spatial patches of the weak heat island areas and the extremely strong heat island areas are small and scattered, whereas the relatively strong heat island areas are characterized by contiguous and concentrated patch distributions.
3. Spatial Association Pattern: The adjacent heat island intensity areas exhibit a "gradient decrease" spatial distribution rule. Within the high-temperature areas, patches with higher temperatures are aggregated, while in the low-temperature areas, patches exhibit diversified temperature ranges.

This comprehensive analysis provides a detailed understanding of the spatial patterns and distribution rules of heat island intensity in Wuhan, highlighting the dynamic interactions and morphological features of heat island areas.

References

1. Deng S et al (2014) ENVI remote sensing image processing methods, vol 7. Higher Education Press, Beijing, pp 2–281
2. Sobrino JA, Raissouni N, Li ZL (2001) A comparative study of land surface emissivity retrieval from NOAA data. Remote Sens Environ 75(2):256–266
3. Ye CH, Liu YH, Liu WD, Liu C, Quan WJ (2011) Research and application of remote sensing monitoring indicators for urban surface thermal environment. Meteorol Sci Technol 39(1):95–101

Chapter 3
Key Urban Form Indicators Affecting Heat Island Intensity

In order to identify the core urban form indicators that affect heat island intensity, this chapter begins by constructing a comprehensive system of potential urban form indicators and detailing the calculation methods for each indicator. This system encompasses four major types of urban form indicators: land use function, land cover, building intensity, and building morphology. By analyzing the spatial heterogeneity of these indicators, it is demonstrated that their spatial variation aligns well with the overall spatial pattern of the heat island effect, thereby validating the hypothesis that these urban form indicators indeed significantly influence urban heat islands. Following this, the chapter proceeds to select the key urban form indicators that most strongly influence heat island intensity. This selection is based on a detailed examination of the urban form characteristics of typical heat island areas, coupled with a comparative analysis of these indicators across different heat island intensity zones. Through this approach, the chapter identifies key urban form indicators that have a substantial impact on the heat island effect, offering insights into the underlying mechanisms of heat island intensity.

3.1 Selection and Calculation Method of Urban Form Indicators

3.1.1 Indicator Selection

There are many types of urban form indicators, but not all of these indicators have a substantial impact on the heat island effect. Therefore, before exploring the relationship between urban form characteristics and heat island intensity, it is essential to select a subset of these indicators as potential factors affecting the heat island effect.

Y. Lu, *Urban Form and Urban Heat Island*, Urban Sustainability,
https://doi.org/10.1007/978-981-96-3541-2_3

This selection process is informed by a comprehensive literature review and an analysis of the functional characteristics of different heat island areas [1]. It is important to clarify that the aim of this section is to propose potential urban form indicators that are hypothesized to influence heat island intensity. However, the validity of these hypotheses will be subject to further empirical testing.

From the existing literature, four key types of urban form indicators that impact the heat island effect can be identified: land use function, land cover, building intensity, and building morphology (see Sect. 2.4). The influence of land use function and building intensity on the heat island effect is primarily related to the heat dissipation from air conditioning systems in buildings. Land cover, on the other hand, affects the local thermal environment through the evapotranspiration cooling effect of vegetation and the radiant heat effect of impervious surfaces on solar radiation. Building morphology influences heat island intensity by modifying the ventilation performance of urban blocks and the shading effects of buildings. Consequently, this book posits that these four types of urban form indicators—land use function, land cover, building intensity, and building morphology—are potential determinants of heat island intensity. Based on these categories, a total of 19 specific urban form indicators are proposed and discussed (see Table 3.1).

3.1.2 Calculation Method

Among the urban form indicators listed above, some are common metrics, such as the proportion of various types of construction land, floor area ratio, and building density, which are frequently used in general urban planning and management. Conversely, other indicators are relatively rare in typical planning practices but may have significant potential for influencing the heat island effect. While the common indicators are well-established with standard calculation methods, the less common indicators are explored here for their potential impact on the heat island effect, with the goal of assessing whether these indicators can effectively reflect the intensity of heat islands.

3.1.2.1 Calculation Method of NDVI

The Normalized Vegetation Index (hereinafter referred to as "NDVI") and the proportion of different types of surface coverage are used to reflect the surface coverage characteristics of each typical heat island area, including the proportion of buildings and roads, the proportion of hardened surfaces, the proportion of vegetation, the proportion of bare surfaces, and the proportion of water bodies.

The NDVI indicator was calculated from the band of the Landsat 8 remote sensing data on October 6, 2014, and the calculation formula is as follows:

$$\mathrm{NDVI} = (\mathrm{NIR} - \mathrm{RED})/(\mathrm{NIR} + \mathrm{RED})$$

Table 3.1 Potential urban form indicators affecting heat island intensity

	Name of index	Unit	Computational method
Functional development indicators	The proportion of industrial land	%	Total area of industrial land (M)/area of the management unit
	The proportion of commercial land	%	Total area of commercial land (B)/ area of the management unit
	The proportion of residential land	%	Total area of the residential land (R)/ the area of the management unit
	Density of permanent population	Thousands of people / km^2	On the basis of "six pu" calculation
	Job density	Ten thousand/km^2	According to the land the calculation method is detailed in Sect. 3.5.1
Surface cover class indicators	Ratio of green space	%	Total amount of green land (G, E2, E9)/area of the management unit
	Normalized vegetation coverage index (Ndvi)		Average observed value of all NDVI measurement points within the management unit
	Building coverage ratio	%	The results was calculated according to the supervised classification as described in Sect. 4.2.3
	The proportion of road coverage	%	
	Hardened surface cover ratio	%	
	The proportion of vegetation coverage	%	
	The proportion of covered soil	%	
	Water body coverage ratio	%	Total amount of water type land (E1)/area of the management unit
Construction intensity category indicators	Plot ratio		Total floor area/ location of the management unit area

(continued)

Table 3.1 (continued)

	Name of index	Unit	Computational method
	Building density		Total area of building base/area of management unit
	The proportion of construction land		The sum of the area of various urban construction land (except green land)/ the area of the management unit
Building combination class index	Harshness		The calculation method is detailed in the text
	Surface sky visibility (hereinafter "GSVF")		
	Building body coefficient		

where, NIR is the near-infrared band, and RED is the red light band. Generally speaking, the higher the NDVI value, the better the greening level.

3.1.2.2 Composition of Different Land Cover

The method of supervised classification is employed to identify different land covers, consisting of two main stages. First, training samples are defined by visually interpreting typical areas of five object types—buildings and roads, hardened surfaces, vegetation, exposed surfaces, and water bodies—within the main urban area of Wuhan. Regions of Interest (ROIs) are established for each type, ensuring their uniform distribution across the entire research area. Second, supervised classification is executed using ENVI software, where the ROI separability tool is utilized to calculate the statistical distance between categories. The "minimum distance" classifier is then applied to perform the classification. This involves calculating the mean vector and standard deviation vector for each category, with the mean vector serving as the central reference point in feature space. Pixels in the input image are assigned to the category whose center is closest in distance, based on the computed mean vectors.

3.1.2.3 Calculation Method of GSVF

GSVF reflects the sky visibility of each point on the surface. The GSVF value for a specific location is obtained by averaging the GSVF values of all points within its range. The basic principle of calculating this indicator is to select a point on the surface as the center, create a hemisphere with a specific radius, and calculate the

Fig. 3.1 Residential area proportion of different regulation planning unit

proportion of the projection area of buildings within this hemisphere relative to the total surface area of the hemisphere (Fig. 3.1). This calculation can be performed using SAGA-GIS software [2, 3]. In this book, the GSVF measurement employs a search radius of 200 m and collects data at a resolution of 10 × 10 m. After importing the building vector data, the GSVF value for each point is computed, and the average GSVF value of all points within each management unit is used as the GSVF indicator for that unit. A higher GSVF value indicates greater openness of the three-dimensional space around the observation point to the sky.

3.1.2.4 Calculation Method of Roughness

Roughness affects air circulation and is an important factor influencing wind circulation potential. The calculation of this indicator must consider not only the three-dimensional configuration of the building group but also the local wind direction. The larger the vertical projection area of the building group in the windward direction, the greater the surface roughness. The calculation formula of surface roughness is as follows:

$$\beta_{f(\theta)} = \frac{A_F}{A_T} = L_y Z_H \rho_{el}$$

where A F represents the windward area of the building, with a wind angle of θ, A T represents the area of the region, Ly represents the average width of the windward roughness object, Z H represents the average height of the building, and ρ el represents the number of buildings per unit area (number density). The windward area refers to the projected area of the building in a certain wind direction, which approximately represents the size of the wind-blocking surface of the building. This book selected the north-east 22.5°, which is the dominant wind direction in Wuhan city, as the basis for calculating the roughness of the non-river typical heat island area, including the Zhuankou Industrial Zone, Hanxi Building Materials Market, Fengsheng Road District, Xingye Road District. And took the direction perpendicular to the Yangtze River as the basic wind direction for calculating the other two typical heat island areas along the river, including the Minquan Road District and Wuhan Tiandi District.

The vertical grid structure of the Urban Canopy Parameter (UCP) reveals the architectural configuration within the urban canopy by analyzing vertical level divisions. This structure reflects the distribution characteristics of building heights in the area and provides a detailed description of the non-uniformity of the urban ground cover. The calculation method involves statistically analyzing building density on horizontal cross-sections at various heights. Specifically, this book uses a gradient of 5 m to analyze the urban canopy structure across different height levels up to 100 m.

3.1.2.5 Calculation Method of Building Morphology Coefficient

The building morphology coefficient equals the ratio of the total external surface area of all buildings in this area in contact with the outdoor atmosphere to the volume they enclose. The building volume equals the area of the building footprint multiplied by the building height, and the surface area is the base perimeter multiplied by the height plus the roof area.

3.2 Spatial Variation of Urban Form Indicators

3.2.1 Spatial Variation of Functional Development and Construction Combination Indicators on Urban–Rural Gradient

3.2.1.1 Spatial Variation of Functional Development Characteristics on Urban–Rural Gradient

The proportion of major construction land types (residential, commercial, industrial) shows a gradual variation from the urban core area to the urban fringe area. In detail, the commercial and residential functions concentrate on the urban core area. Meanwhile, the industry area lies in the outer area. With the increase of the circle distance, the proportion of residential land, commercial land, and industrial land showed relatively regular changes (Figs. 3.3, 3.4 and 3.5). Firstly, the proportion of residential land is higher in the city center with a closer circle distance, and it decreases steadily as the circle distance exceeds 5000 m, with slight fluctuations near the 20,000-m circle distance, indicating significant ring differentiation characteristics of residential land. Secondly, the proportion of commercial land fluctuates considerably across different circles, but overall decreases with increasing distance from the city center. Within a circle distance of less than 5000 m, the proportion of commercial land is generally higher, and it falls below 2.00% beyond 15,000 m. Lastly, the proportion of industrial land exhibits an opposite trend to that of residential and commercial land; it increases significantly with greater circle distances, particularly after 10,000 m, where the proportion of industrial land consistently rises.

Within each heat island intensity area, the trends and degrees of change in functional development elements vary in line with buffer distance. In the extreme heat island area and the strong heat island area, as the distance from the city center increases, the proportion of residential land gradually decreases. On the contrary, the proportion of industrial land increases significantly. Beyond 12,000 m from the city center, the proportion of industrial land surpasses that of residential land, a trend that is more pronounced in the extreme heat island area. Conversely, in the strong heat island area and the weak heat island area, the proportions of residential and industrial land remain relatively stable across different locations.

In different locations, the changes in functional elements with varying heat island intensities are distinct. For residential land, the proportion in the core area of the extreme and strong heat island regions is very high and decreases with increasing circle distance, indicating that residential land is a significant factor in the formation of extreme and strong heat island areas, though this effect diminishes in the outer circles. Commercial land shows a similar spatial pattern to residential land, with a higher proportion closer to the city center and a marked decrease beyond the 12,000–14,000 m circle. Industrial land, on the other hand, increases sharply with distance from the city center, peaking at 64.73% in the 14,000–16,000 m circle in the extreme heat island area, while remaining relatively stable in the strong and weak heat island areas.

Thus, functional elements related to heat island intensity vary with different locations. In the layers closer to the city center (within the buffer distance of 6000 m), higher heat island intensity areas are characterized by a high proportion of residential and commercial land use, which decreases as the distance from the center increases. Conversely, in the layers farther from the city center (beyond 12,000 m), the proportion of industrial land use increases rapidly in areas of extremely strong and relatively strong heat islands. This observation aligns with the spatial differentiation characteristics of the "high inside and outside" heat island intensity pattern described earlier, indicating that different functional elements contribute to the formation of high-intensity heat island areas in varying layers.

After analyzing the proportions of land area and building area for residential, commercial, and industrial functions in each management unit, it is evident that residential functions are widely distributed, particularly dense along the riverfront and the vertical river axis in Hankou, while Hanyang shows less residential distribution outside the Han River area (Fig. 3.1). The residential building area concentration within the first ring does not exhibit a significant advantage, with a more pronounced residential function around large lakes and some peripheral areas (Fig. 3.2). The proportion of commercial functions is higher within the first ring, decaying slightly beyond it, with a uniform distribution between the first and second rings and a rapid decline outside the second ring, reflecting a typical monocentric aggregation (Figs. 3.3 and 3.4). Industrial functions are concentrated in several industrial areas around the third ring, with minimal distribution within the second ring (Figs. 3.5 and 3.6). The permanent population and employment positions exhibit single-center and multi-center spatial distribution patterns, respectively, highlighting the "living inside and producing outside" relationship (Figs. 3.7 and 3.8). The population density is extremely high in the riverfront area within the first ring, peaking at 103,400 people/km^2, while it remains high in main road areas within the second ring but decreases overall. Outside the second ring, only a few areas in the northeast and southeast maintain higher population densities, with a general further decrease. Employment positions show a uniform and dispersed distribution, with medium-density uniformity within the second ring and peaks in several large educational and research areas beyond the second ring, reflecting the concentration of industrial functions around the third ring.

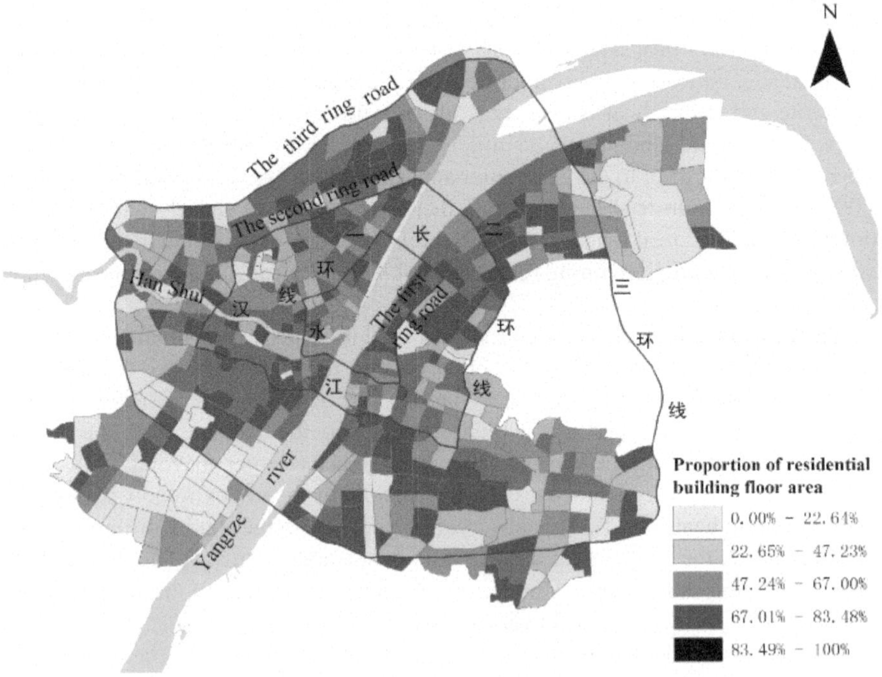

Fig. 3.2 Residential floor area proportion of different regulation planning unit

Fig. 3.3 Commercial area proportion of different regulation planning unit

It can be seen that some functional development indicators within the research scope show a gradual variation along the urban–rural direction. This study posits that functional development is a key spatial morphology factor affecting heat island intensity, where variations in functional elements influence the heat island intensity and contribute to the observed layer differentiation pattern.

3.2.1.2 Spatial Variation of Building Combination Characteristics on Urban–Rural Gradient

The roughness, GSVF, and building form factor of each management unit were calculated separately, revealing regular changes in these variables with respect to layering, demonstrating a strong correspondence among them. In the core area, the dense building groups result in higher roughness values and lower GSVF and building form factor values (Figs. 3.9 and 3.10). Conversely, in the peripheral areas, roughness is lower, while the building form factor and GSVF values are higher. It means that the wind resistance in the core area is large and the space openness is poor, while the wind resistance in the outer area is small and the space is open. At the same time, the overall spatial pattern of the building form coefficient is “low inside and high

Fig. 3.4 Commercial floor area proportion of different regulation planning unit

outside", which means that the building form in the core area is more regular, while in the periphery is relatively scattered.

3.2.2 Spatial Pattern of Land Cover Indicators

Using the correlation analysis for identifying surface cover outlined in Sect. 4.1.2, the spatial distribution of various surface cover class indicators within the overall research scope was calculated separately (Fig. 3.11). The statistical results reveal that different land cover types exhibit significantly different spatial distributions. Generally, the spatial distribution patterns of building cover and road cover show a decreasing trend from the city center to the periphery, although their distribution patterns exhibit slight differences. Specifically, areas with a higher proportion of building cover are predominantly axial, concentrated in local areas such as both sides of the Yangtze River, both sides of the Han River, and the Wuhan-Luoyu Road axis (Fig. 3.12). In contrast, areas with a larger proportion of road cover are concentrated in two distinct fan-shaped regions: one extending northwest from the Hankou riverfront core area and another extending south from the Wuchang riverfront core area (Fig. 3.13). The spatial distributions of soil cover, vegetation cover, and

Fig. 3.5 Industrial area proportion of different regulation planning unit

NDVI values are similar, showing high aggregation in local areas on the periphery of the research scope (Figs. 3.14, 3.15 and 3.16). The waterbody cover shows no significant correlation with the distance from the circle layers and occupies a larger proportion in areas with large-scale lakes (Fig. 3.17). Hardened surface cover is primarily distributed outside the second ring in Hankou and Hanyang and is rarely found within the second ring.

In summary, artificial surface covers are mainly concentrated in the core circle of the city and along the main development axes, whereas ecological surface covers are predominantly found at the edges of the main city and around large-scale water bodies.

3.2.3 Multi-axial Extension of Construction Intensity Class Indicators

The building density, floor area ratio, and proportion of construction land for each management unit were calculated separately (Figs. 3.18, 3.19 and 3.20). Overall, the Hankou riverfront area in the center exhibits extremely high construction intensity, which gradually decreases moving outward from the center but shows local increases

Fig. 3.6 Industrial floor area proportion of different regulation planning unit

in construction intensity at the periphery. These three indicators all demonstrate high values along major city roads and expressways, including the belt areas along the Yangtze River in Wuchang, Hankou, and Hanyang, as well as the belt areas on both sides of the Han River. This indicates that construction intensity features an axial spatial distribution pattern. Notably, the areas with higher construction intensity align closely with regions of higher heat island intensity along the main city roads, revealing a significant similarity in the urban form distribution structures of these two phenomena.

Fig. 3.7 Population density of different regulation planning units

3.3 Spatial Morphological Characteristics of Typical Heat Island Areas

3.3.1 Selection of Typical Heat Island Areas

3.3.1.1 Selection Method

The purpose of studying typical heat island areas is to identify the core urban form indicators that affect the heat island effect. However, the previous analysis has demonstrated that similar heat island effects can emerge under different functional development characteristics. For example, extremely strong heat island areas can include not only concentrated industrial zones such as Wugang and Zhuankou but also densely populated city center areas. This observation highlights that different urban form elements might produce similar heat island effects through various formation mechanisms, making it challenging to derive universal laws governing these effects. To address this complexity, it is essential that the selection of typical heat island areas reflects a range of representative "situations" that capture diverse functional development characteristics and exhibit significantly varied heat island intensities.

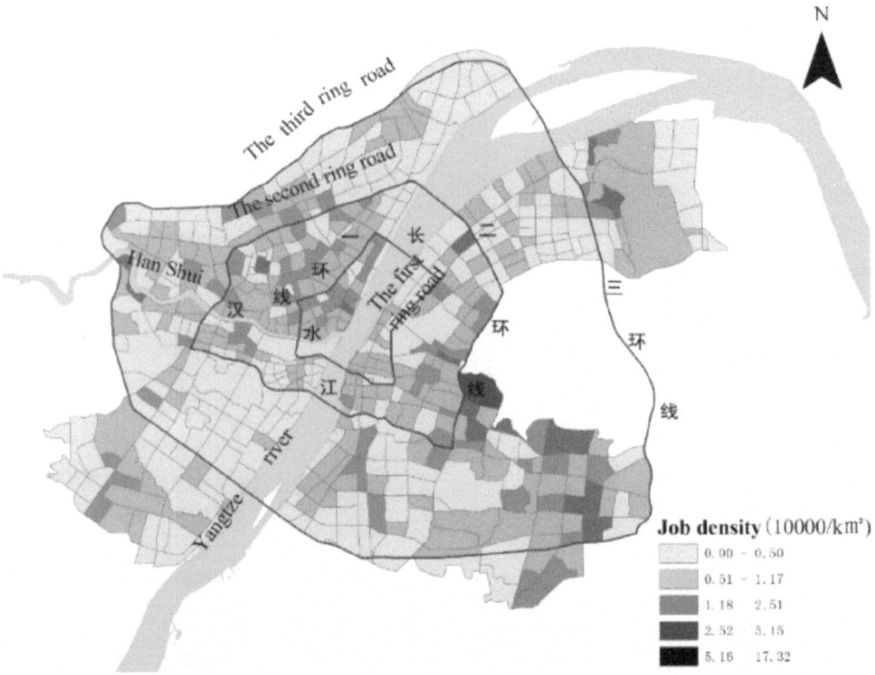

Fig. 3.8 Job density of different regulation planning units

Therefore, the selection of typical heat island areas must be guided by two key principles. First, it is crucial to ensure the typicality of heat island intensity by choosing areas that exhibit stable and extreme heat island intensities, thereby establishing a clear causal relationship between urban form characteristics and heat island intensity. Second, the selected areas should represent diverse functional development characteristics, meaning they should reflect specific urban forms that lead to either extremely high or low heat island intensities. By adhering to these principles, the results of the indicator screening will be more comprehensive and the influencing factors identified will possess greater explanatory power.

The selection method of typical heat island areas is as follows:

- Step One: Extract the land surface temperatures of Landsat 8 for October 6, 2014, and March 31, 2015. Overlay them with 500 × 500 m spatial grid vector data, and calculate the average heat island intensity for each grid cell. This process uses the spatial grid as the analytical unit to determine the average heat island intensity for both dates (Fig. 3.21).
- Step Two: After extracting the extremely strong heat island areas and weak heat island areas for the two time periods, we classify the spatial grids as follows: those that are identified as extremely strong heat island areas at both time points are designated as "stable extremely strong heat island areas", while those identified

Fig. 3.9 Roughness of different regulation planning unit

as weak heat island areas at both time points are designated as "stable weak heat island areas". This Chapter identified 402 spatial grids as extremely strong heat island areas from the image of October 6, 2014. For the image from March 31, 2015, the top 402 spatial grids with the highest heat island intensity are selected as the corresponding extremely strong heat island areas. The selection process for weak heat island areas follows a similar method.

- Step Three: Collect functional development indicators for each spatial grid to serve as variables for cluster analysis. These indicators include the proportion of residential land, commercial land, industrial land, green land, public management, and public service land, as well as permanent population density and employment position density. Systematic clustering is performed on the spatial grids from both the stable extremely strong heat island areas and the stable weak heat island areas using SPSS software. This analysis results in the identification of three types of stable extremely strong heat island areas and three types of stable weak heat island areas, with each type representing a regional space with distinctive functional development characteristics.
- Step Four: To ensure consistency in heat island intensity within each case, areas with stable heat island intensity are selected based on the Shannon diversity statistical results discussed in Sect. 3.2.3. This step involves selecting regions where the heat island intensity remains consistent and can be linked to urban road and

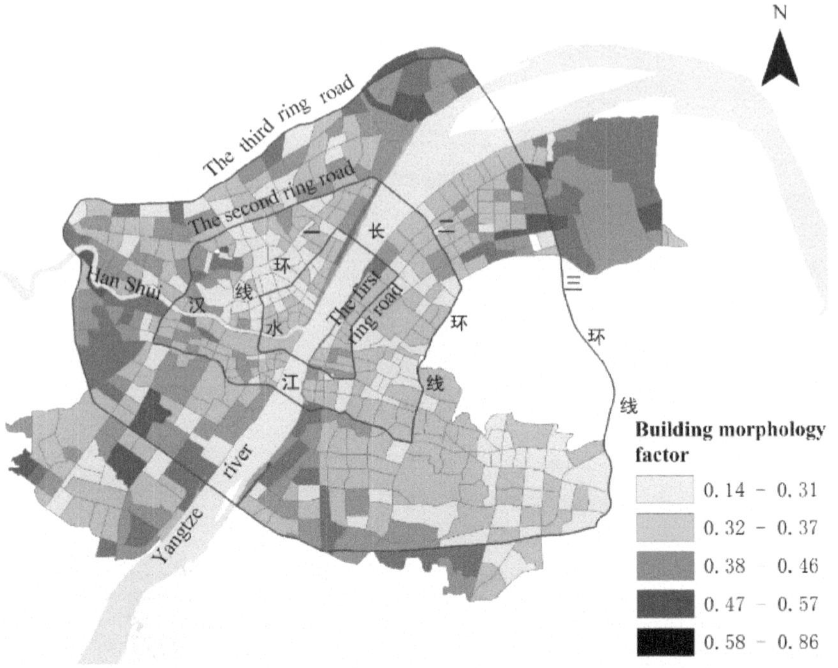

Fig. 3.10 Building morphology factor of different regulation planning unit

building mechanisms. A typical heat island area is then chosen from each type of spatial grid according to the clustering results from Step Three, ensuring that the selected areas serve as representative case studies for the different heat island intensity types.

3.3.1.2 Selection Results of Typical Blocks

After intersecting the extreme heat island areas from October 6, 2014, and March 31, 2015, a total of 286 spatial grids were identified as stable extreme heat island areas (Fig. 3.21). Using the WARD algorithm in SPSS software, these grids were systematically clustered based on their functional development features, resulting in the identification of three major types of stable extreme heat island areas: "low-density industrial type," "high-density service type," and "high-density residential type" (Fig. 3.22).

The functional parameters of each type obtained from the clustering analysis were statistically examined, revealing significant differences among them. The "density" categorization is based on a summary of permanent population density and employment density, while the dominant functional types are characterized by their specific land use function features.

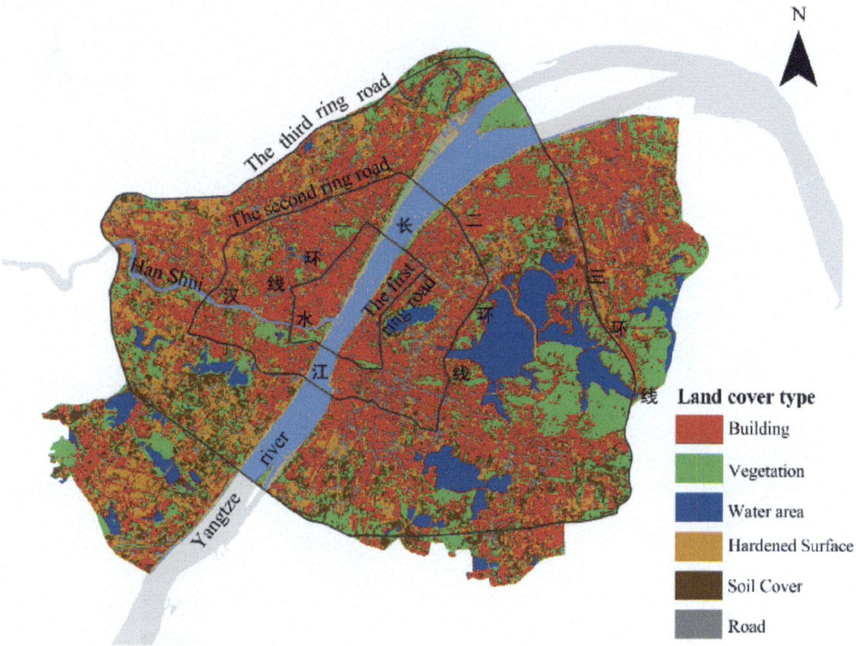

Fig. 3.11 Land cover type

- Low-Density Industrial Type: In this category, industrial land accounts for a substantial 61.00% of the total land area. The permanent population and employment densities are the lowest among the three types, at only 2600 people/km^2 and 6800/km^2, respectively. This indicates that industrial functions are dominant in this kind of block, with minimal presence of other urban functions.
- High-Density Service Type: This type is characterized by significant commercial and public service functions, with land use proportions reaching 21.07% for commercial services and 30.36% for public services. The employment density of this kind of block is the highest among the three types, reflecting a concentration of intensive service functions.
- High-Density Residential Type: Residential land in this type covers 40.36% of the area, often mixed with some commercial land. The permanent population density is exceptionally high, reaching 2800 people/km^2, which categorizes this area as a high-density residential zone.

The same method was used for clustering analysis of the spatial grids in the stable weak heat island area, and according to their functional parameters, they were also divided into three categories, namely "low-density ecological type", "medium-density residential type" and "high-density mixed commercial and residential type". In the "low-density ecological type" spatial grid, the proportion of construction land and the density of permanent population and employment positions are very low,

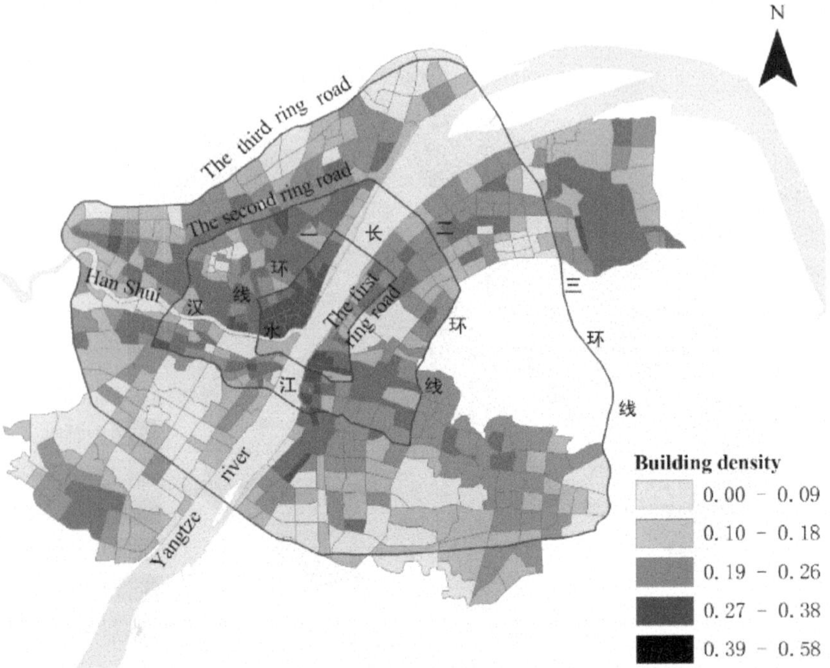

Fig. 3.12 Building cover of different regulation planning unit

indicating that this type is dominated by non-construction land, and the urban function is not fully developed. The residential land in the "medium density residential type" spatial grid is concentrated and contiguous, accounting for 59.61%, but the functional development of other cities is insufficient, among which the commercial land accounts for only 1.24%. The "high-density commercial and residential" type has a small number of spatial grids and a very high resident population density.

Based on the clustering results from the stable extreme heat island areas and stable weak heat island areas, as well as the functional development characteristics of each type, we select one block from each of the six types as representative cases (Fig. 3.23). Specifically, the Chuankou Industrial Zone in the Wuhan Economic and Technological Development Zone, the Hanxi Building Materials Market in Qiaokou District, and the Minquan Road neighborhood in the old city of Hankou are selected as representative cases of "low-density industrial type," "high-density service type," and "high-density residential type," respectively (Fig. 3.24). For the stable weak heat island areas, the Fengsheng Road neighborhood in Hongshan District, Xingye Road neighborhood in Jiang'an District, and Wuhan Tiandi neighborhood in Jiang'an District are selected as representative cases for "low-density ecological type," "medium-density residential type," and "high-density commercial-residential mixed type," respectively (Fig. 3.25). The functional indicators of each selected heat island area closely match the characteristics of their respective types,

Fig. 3.13 Road surface cover of different regulation planning unit

and the locations align with the functional spatial analysis results for extreme and weak heat island areas as discussed in Sect. 3.3.1. Specifically, the extreme heat island areas in the core and peripheral layers are characterized by high-density residential and industrial functions. While the weak heat island areas are primarily characterized by lower-density residential functions and green spaces.

3.3.2 Spatial Characteristics of Typical Extreme Heat Island Areas

3.3.2.1 Functional Development Characteristics: Concentrated Industrial, Commercial Areas or High-Density Residential Areas

The land use function characteristics of each typical heat island area align closely with the functional classifications derived from the clustering results (Table 3.2). Examining the two groups of typical heat island areas with vastly different heat island intensities reveals that residential areas with extremely high population density have a significant potential to form extreme heat island areas. Among the six typical heat

Fig. 3.14 Vegetation cover of different regulation planning unit

island areas selected in this study, three are predominantly residential: Minquan Road neighborhood, which is classified as a stable extreme heat island area, and Xingye Road neighborhood and Wuhan Tiandi neighborhood, both of which are classified as stable weak heat island areas. This observation suggests that the correlation between residential function type and heat island intensity is not strongly significant. However, the intensity of the residential function plays a crucial role. For instance, Minquan Road neighborhood, with an exceptionally high permanent population density of 87,100 people/km^2, exemplifies how residential functions can contribute to extreme heat island intensity only when they reach such an extreme level of density.

In contrast, the other two typical heat island areas classified as stable extreme heat island areas are characterized by dominant industrial and commercial land uses. These areas exhibit a medium level of employment density, with industrial and commercial land uses comprising 92.72% and 96.47% of their respective areas. This indicates that, when industrial and commercial functions reach a certain scale and become relatively concentrated, they can also exhibit extremely high heat island intensity.

Fig. 3.15 NDVI of different regulation planning unit

3.3.2.2 Land Surface Cover Characteristics: Dominated by Impervious Surface, Lack of Ecological Surface

According to the composition of land surface coverage, the three block cases belonging to the stable extreme heat island area are dominated by buildings and roads, as well as hardened surfaces. The combined proportion of these two types of impervious surfaces reaches 93.13%, 100.00%, and 99.43% respectively. Among them, the impervious surface area in the Zhuankou Industrial block and the Hanxi Building Materials Market has a higher proportion, reaching 56.50% and 41.57% respectively, while the Minquan Road block is covered by buildings and roads, with a proportion as high as 94.82%. Vegetation coverage and soil coverage are very small in these three cases. Compared with the overall surface coverage of the research scope, the typical heat island areas belonging to the stable extreme heat island area have obvious advantages in the building, road, and impervious areas (Table 3.3).

The NDVI value and the green space rate also show that the ecological level of the three cases in the stable extreme heat island area is poor (Fig. 3.26). The green space rates of Zhuankou Industrial Area, Hanxi Building Materials Market, and Minquan Road block are 1.39%, 0%, and 0% respectively, and the NDVI value hovers at a low level, with the highest value of only 0.11, significantly different from the average

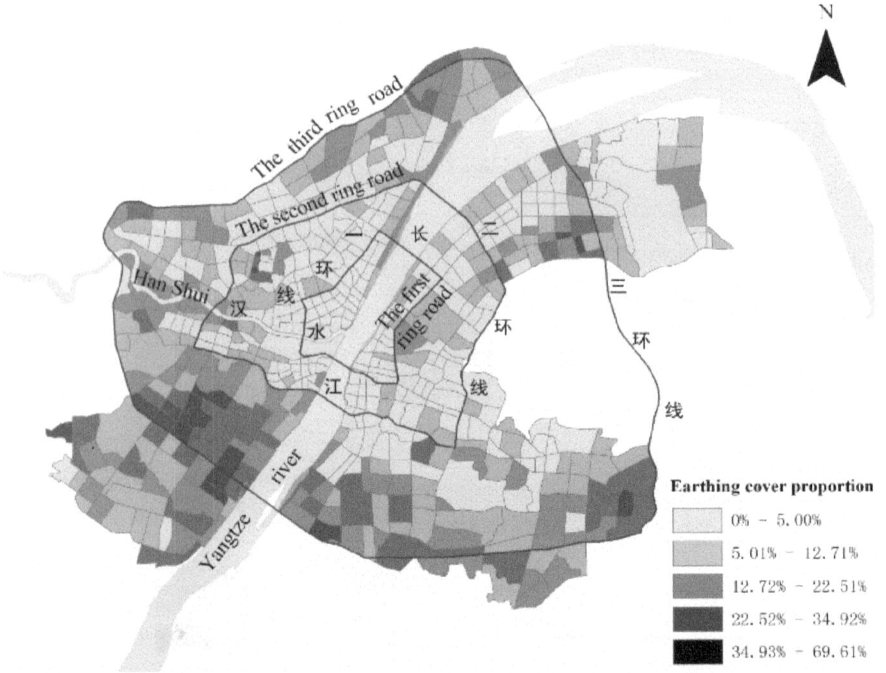

Fig. 3.16 Earthing cover of different regulation planning unit

level within the research scope and the other three cases belonging to the stable weak heat island area.

In summary, the common characteristics of the stable extreme heat island area are the dominance of impermeable surfaces and the lack of ecology.

3.3.2.3 Construction Intensity Characteristics: High Building Density

Building density, floor area ratio, and average number of floors are used to measure construction intensity characteristics. The three typical heat island areas classified as stable extreme heat island areas exhibit notable differences in their floor area ratio. Among these, the Zhuankou Industrial Area has the lowest floor area ratio at 0.52, whereas the Minquan Road block features the highest floor area ratio at 2.23. Despite these differences, all three areas demonstrate high building density, which is significantly greater than that observed in other regions. The average number of floors for the Zhuankou Industrial Area, Hanxi Building Materials Market, and Minquan Road block are 1.46, 2.44, and 3.67, respectively. While the first two cases show a higher building density compared to other functional areas, the Minquan Road block, dominated by residential functions, has a lower average number of floors and a higher building density compared to other residential areas.

Fig. 3.17 Water cover of different regulation planning unit

In summary, the common characteristics of stable extreme heat island areas are high building coverage and a lower average number of floors. Whereas the floor area ratio does not exhibit a direct relationship with the formation of stable extreme heat island areas.

3.3.2.4 Building Combination Characteristics: Compact, Aggregated, Low Density and High Sparsity

Typical heat island areas classified as stable extreme heat island areas exhibit both differential and common characteristics in their building combinations (Fig. 3.27).

The building combination characteristics of the Zhuankou Industrial Area and the Hanxi Building Materials Market are quite similar. The average building footprint area in these two areas ranks highest among all typical heat island areas, reaching 1164.10 m^2 and 707.13 m^2 respectively (Table 3.4), which results in a wider windward width. However, compared to Zhuankou Industrial Area, the Hanxi Building Materials Market has greater roughness and a lower GSVF index, indicating that the building spacing is smaller and the building height is higher, thus resulting in a more compact building arrangement. In contrast, the building combination form of the Minquan Road block is significantly different from the previous two areas. Here,

Fig. 3.18 Building density of different regulation planning units

the average base area of the building unit and the GSVF are considerably lower, while the roughness, building form factor, and GSVF are higher, reflecting a more complex and variable single building form. Although the average windward surface is smaller in Minquan Road block, the dense grouping of buildings creates a more compact and enclosed space.

Despite these significant differences in building combination space indicators among the three blocks, there are notable common features. First, the high building density results in extremely compact buildings, with Hanxi Building Materials Market and Minquan Road block exhibiting such intense building concentration that open spaces are nearly nonexistent. Second, the two-dimensional spatial distribution of buildings exhibits large-scale block aggregation; for instance, the Zhuankou Industrial Area and Hanxi Building Materials Market have large average building base areas, while the Minquan Road block, despite having smaller individual building base areas, features a high density of small-scale buildings that collectively form a block-shaped building group. Third, all three cases display a similar vertical structure in their UCP grid, characterized by high building density at lower canopy heights and a sharp decline in building density as the canopy height increases, demonstrating a "low density at the bottom and high-density at the top" pattern in the building combination form.

Fig. 3.19 Built-up area proportion of different regulation planning unit

3.3.3 Spatial Features of Typical Weak Heat Island Areas

3.3.3.1 Functional Development Features: Sparse Construction Areas, Modern Residential Function Dominated Single Residential Areas or Mixed Commercial and Residential Complexes

Urban construction in sparse areas is generally less likely to produce strong heat island effects. For instance, the Fengsheng Road block is characterized predominantly by agricultural and forestry land, which constitutes 81.19% of the area, while construction land makes up only 11.11%. Additionally, both permanent resident density and employment position density in this area are nearly negligible. In contrast, the three cases identified as stable extreme heat island areas exhibit a nearly complete dominance of urban construction land, approaching 100%. This stark contrast illustrates that a high proportion of urban construction land is a crucial prerequisite for the development of a strong heat island effect, while sparse urban functions and low-density land uses are insufficient for generating significant heat accumulation.

The second and third types of stable weak heat island areas are represented by the Daxing Road block and the Wuhan Tiandi block, which both feature functionally uniform, modern residential areas. The Daxing Road block encompasses four distinct residential communities—Xiangli Guoting, Xiangshan Garden, Dongfang Shijia,

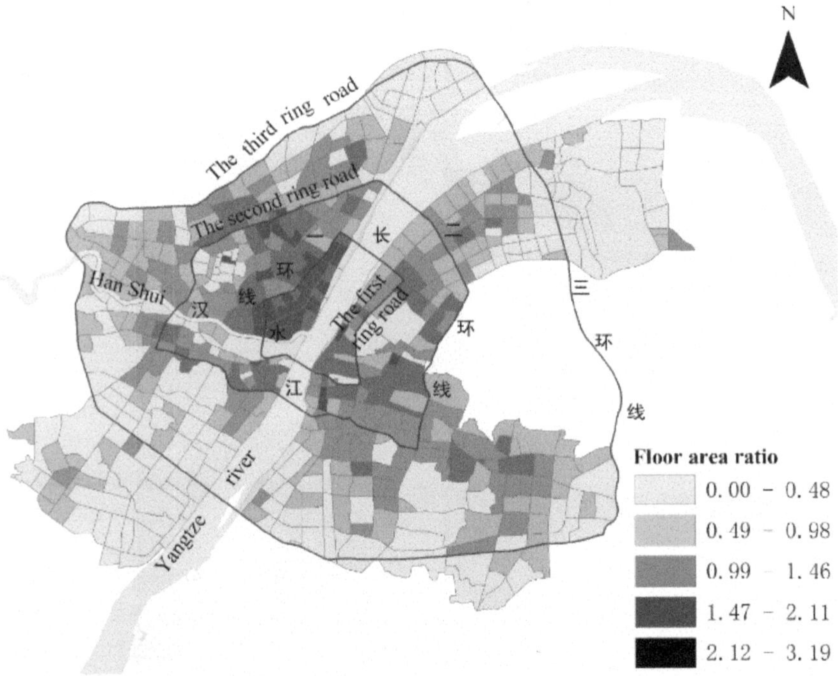

Fig. 3.20 Floor area ratio of different regulation planning unit

and Dongfang Mingzhu—that were developed and launched in June 2007, January 2006, January 2006, and January 2007, respectively. The Wuhan Tiandi block, a mixed-use commercial and residential complex, includes Changyang Yongqing City, Yujiangyuan Phase I, and Wuhan Tiandi Business Group, which were launched for sale in 2009. The Daxing Road block is a highly homogeneous residential area with a residential land ratio of up to 98.96%, whereas the Wuhan Tiandi block features a more mixed land use, with residential and commercial lands comprising 54.21% and 30.98% of the total area, respectively. These residential communities exemplify modern developments that adhere to contemporary technical specifications, in sharp contrast to the high-density, older residential areas found in typical strong heat island regions.

3.3.3.2 Surface Cover Features: Impervious Surface Mixed with Green Space, Abundant Ecological Elements

Compared to the three cases in the stable extreme heat island areas, the surface coverage characteristics of the typical heat island areas in the stable weak heat island areas exhibit a diverse range of trends, which can be summarized into two distinct types.

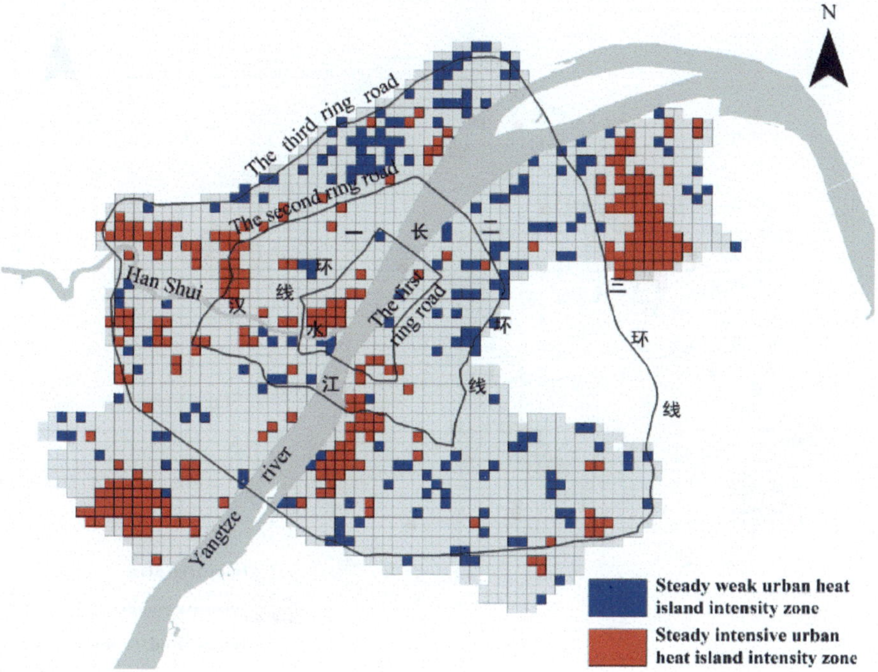

Fig. 3.21 Spatial layout of stable extreme urban heat area and stable weak urban heat area zone

In Fengsheng Road block, due to the incomplete urban functional development, a substantial portion of the area remains undeveloped. Consequently, the surface coverage is predominantly composed of vegetation and covered soil, which account for 67.85% and 15.77% of the total area, respectively. The proportion of impervious surfaces in this block is only 16.38%, and the average NDVI value is relatively high at 0.25 (Fig. 3.28). This indicates a predominantly green and less developed environment.

In contrast, the Xingye Road block and the Wuhan Tiandi block, which are also representative of the stable weak heat island areas, exhibit a different surface coverage profile. In these areas, buildings and roads dominate, with the proportions of these surfaces reaching as high as 95.08% and 86.73% respectively. The NDVI values in these blocks are both around 0.11, reflecting a significantly lower level of vegetation and a more built-up environment.

Overall, the NDVI indicators for the three typical cases in the stable extreme heat island areas are higher compared to those in the three typical cases in the stable weak heat island areas. This suggests that the ecological conditions in the stable extreme heat island areas are markedly worse than those in the stable weak heat island areas. The ecological level of the stable extreme heat island areas is significantly inferior to that of the stable weak heat island areas, highlighting a clear distinction in environmental quality between these two types of heat island areas.

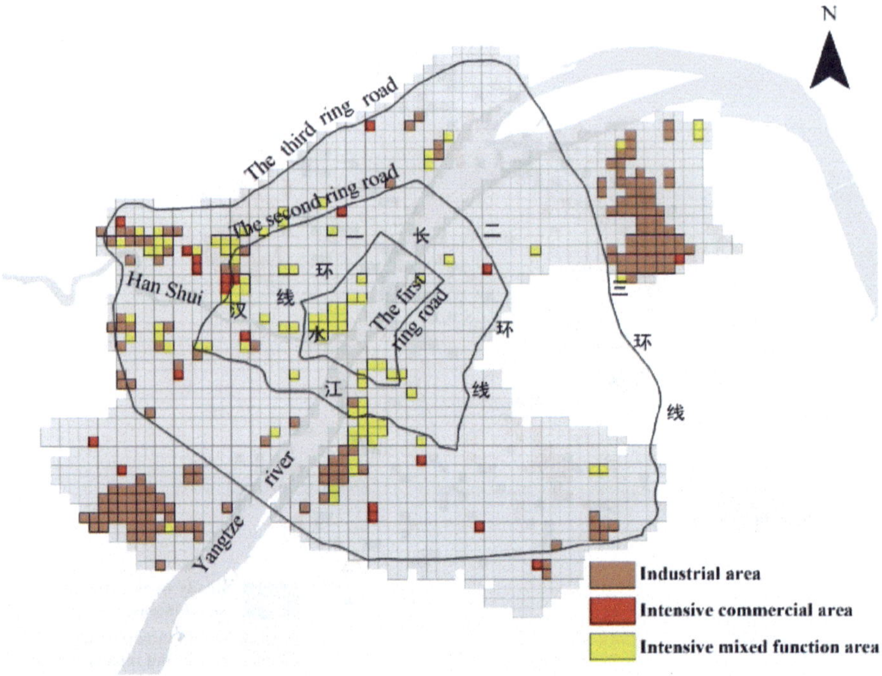

Fig. 3.22 Classification of stable extreme urban heat area zone

3.3.3.3 Construction Intensity Characteristics: Low Building Density

The stable weak heat island areas can be categorized based on their construction intensity indicators into two distinct groups: the Fengsheng Road block, which has an underdeveloped urban function with existing buildings primarily consisting of low-rise farmhouses, resulting in extremely low construction intensity indicators, and the Xingye Road block and Wuhan Tiandi block, which were developed in the past decade and exhibit similar construction intensity indicators. Both areas have low building density but high floor area ratios, distinguishing them significantly from the Fengsheng Road block.

While the Xingye Road block and Wuhan Tiandi block share similarities with the Minquan Road block—a stable strong heat island area dominated by residential functions—they also exhibit notable differences. The floor area ratios of Xingye Road block, Wuhan Tiandi block, and Minquan Road block are similar, with values of 2.02, 1.96, and 2.23, respectively. However, there are significant differences in building density and the average number of floors among these blocks. The stable weak heat island areas (Xingye Road block and Wuhan Tiandi block) exhibit characteristics of "low density, high height," while the Minquan Road block is characterized by "high density, low height." Low building density is a common feature across the stable weak heat island areas, contributing to their lower heat island intensity compared to

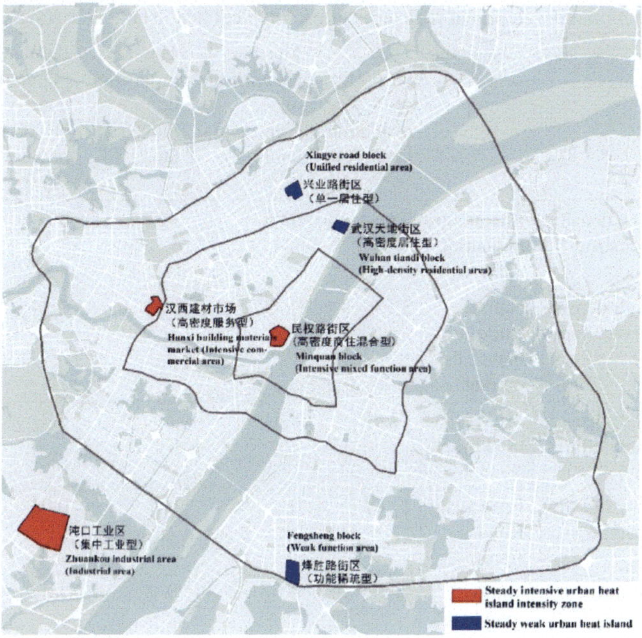

Fig. 3.23 Typical urban heat island

Fig. 3.24 Land use of typical stable extreme urban heat island intensity zone

Fig. 3.25 Land use of typical stable weak urban heat island intensity zone

Table 3.2 Functional development indicators of each typical heat island area zone

		Case name	The proportion of residential land (%)	The proportion of commercial land (%)	The proportion of industrial land (%)	Permanent population density (10,000 people / km^2)	Job density (10,000 people /km^2)
Stabilize the extremely strong heat island zone	Low density industrial type	Zhuankou industrial zone	0.66	0.95	92.72	0.08	1.35
	High density service type	Hanxi Building materials market	3.53	96.47	0	0.54	3.16
	High density residential type	Mingquan road block	77.55	8.51	2.12	8.71	0.62
Stabilize weak hot island areas	Low density ecotype	Fengsheng Road Block	1. 67	0	0	0.03	0
	Medium density residential type	Xingye road block	98.96	0	0	1.47	0
	High-density commercial and residential hybrid type	Wuhan Tiandi block	54.21	30.98	0	1.61	2.34

Table 3.3 Land cover indicators of each typical heat island area zone

	The proportion of building and road (%)	The proportion of hardened surface (%)	The proportion of vegetation (%)	The proportion of coverage ratio (%)	The proportion of water (%)
Zhuankou industrial zone	36.63	56.50	1.28	5.59	0.00
Hanxi building materials market	58.43	41.57	0.00	0.00	0.00
Mingquan road block	94.82	4.61	0.00	0.57	0.00

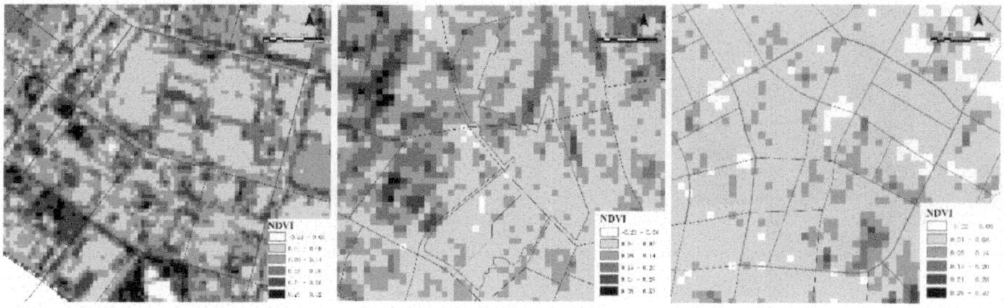

Fig. 3.26 NDVI of table weak urban heat island intensity zone

Fig. 3.27 Building storey of typical heat island area zones

Table 3.4 Building combination characteristic of each typical heat island area zone

	Average windward width	Roughness	Building body coefficient	GSVF	Average building base area (m^2)
Zhuankou industrial zone	36.44	0.0766	4.40	0.9836	1164.10
Hanxi building materials market	33.67	0.1674	6.90	0.9214	707.13
Mingquan road block	17.95	0.3932	2.93	0.8457	220.10
Fengsheng road block	9.99	0.0082	1.72	0.9980	135.75
Xingye road block	25.09	0.3594	3.42	0.7533	2744.13
Wuhan Tiandi block	18.85	0.2217	3.99	0.8175	316.11

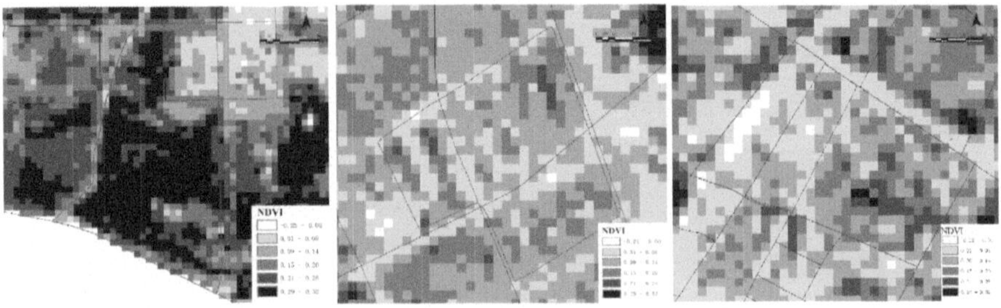

Fig. 3.28 NDVI of stable weak urban heat island intensity zone

areas with higher building density. This suggests that lower building density is a key factor in mitigating the urban heat island effect.

3.3.3.4 Building Combination Characteristics: Loose and Open, Point-Like Dispersion, Uniform Height

The three typical heat island areas in the stable weak heat island zone can be categorized into two types. In the Fengsheng Road block, due to underdeveloped urban functions, buildings are sparsely distributed with large areas dedicated to agriculture and forestry and a few scattered individual buildings. In contrast, the Xingye Road block and Wuhan Tiandi block have densely distributed buildings with distinctive architectural spatial combination characteristics (Fig. 3.29).

These two types of stable weak heat island areas exhibit significant differences in architectural space combinations. The Fengsheng Road block exemplifies a combination dominated by low-density, low-height, low-rise buildings. Meanwhile, the Xingye Road block and Wuhan Tiandi block share similarities, representing combinations characterized by low-density, high-rise buildings. Examining the architectural combination indicators, the majority of buildings in the Fengsheng Road block are not higher than three floors. Consequently, the windward area and roughness are extremely low, and building density remains low at various height levels. The space

Fig. 3.29 Buildings of stable weak urban heat island zone

is open with high sky visibility, indicating that the architectural combination here has minimal impact on air circulation, maintaining excellent ventilation potential and allowing internal heat to dissipate quickly.

Conversely, the Xingye Road block and Wuhan Tiandi block exhibit extremely high roughness (Table 4.4). Although their average windward width is not as extensive as in the Zhuankou Industrial Zone and Hanxi Market Zone, and their building layouts are scattered, the extremely high average building height and the slow rate of density decrease with rising height levels result in more vertical windward area. This leads to high roughness and relatively low GSVF.

In summary, the three cases in the stable weak heat island area share the following characteristics:

1. In the two-dimensional plane combination of buildings, the spacing between each building unit is large, creating sparse building coverage and ample open space.
2. The overall architectural layout of each case exhibits a "point-like scattered" mode, sharply contrasting with the dense arrangements observed in the Zhuankou Industrial Zone and Hanxi Building Materials Market within the stable strong heat island area.
3. From the analysis of the UCP vertical grid structure, the vertical structure in each case is uniform, with building height remaining relatively stable and not decreasing rapidly with increased height, unlike the cases in the stable strong heat island area.

3.4 Comparison of Urban Form Indicators in Different Heat Island Intensity Areas

Through the previous comparison of the urban forms of typical strong heat island areas and typical weak heat island areas, it was found that there are significant differences in functional development, land surface coverage, construction intensity, and building combination. To further identify which urban form indicators are more closely related to the intensity of the heat island effect, this section will quantify the urban form indicators in areas with varying heat island intensities and analyze the relationship between these indicators and the changes in heat island intensity.

3.4.1 Comparison of Functional Development Indicators

According to the previous criteria for classifying heat island intensity areas, regulatory planning units belonging to extremely strong heat island areas, strong heat island areas, relatively strong heat island areas, and weak heat island areas were identified. The functional development indicators for each management unit in these different heat island intensity areas were compiled, and the maximum, third quartile, median, first quartile, and minimum values of each indicator were calculated. These values were visualized using box plots. The analysis revealed that heat island intensity within a management unit is regularly associated with various functional development indicators. There is a proportional relationship between heat island intensity and commercial functions, industrial functions, and the carrying capacity of permanent residents and jobs. Specifically, areas with a higher proportion of commercial and industrial land use, as well as higher population and job densities, tend to exhibit stronger heat island effects.

These statistics indicate that areas with more concentrated and intensively used commercial and industrial functions tend to form stronger heat islands. The proportion of different functional land uses and building areas reflects the functional composition within each management unit, while the density of the permanent population and employment positions reflects the intensity of functional use. The density of the permanent population indicates the intensity of urban residential function use, while the density of employment positions indicates the intensity of commercial and industrial function use.

The relationship between residential function and heat island intensity is complex. Generally, the more pure the residential function, the lower the heat island intensity. However, the greater the intensity of residential function use (manifested as higher permanent population density), the higher the heat island intensity. This phenomenon could be due to the fact that while residential areas generate heat discharge, which accumulates with an increasing permanent population, the unique spatial characteristics of residential areas (such as higher green space ratios and more building shadows) result in a lower overall heat island effect in areas where residential land use is concentrated.

3.4.2 Comparison of Land Surface Cover Indicators

Eight indicators reflecting surface cover were statistically analyzed, revealing that seven of these indicators have a regular relationship with heat island intensity, with the exception of the proportion of road cover. The green space rate, NDVI, soil cover proportion, vegetation cover proportion, and water body cover proportion are inversely proportional to heat island intensity, while the building cover proportion and hardened surface cover proportion are directly proportional to heat island intensity.

The inverse indicators have varying degrees of impact on heat island intensity. The negative effect of water body cover proportion on heat island intensity is the strongest. In weak heat island areas, the average water body cover proportion significantly increases to 11.71%, whereas extremely strong heat island areas have almost no water body cover. The soil cover proportion has a relatively small correlation with heat island intensity, slightly increasing from 5.06% in extremely strong heat island areas to 11.41% in weak heat island areas. The green space rate, NDVI, and vegetation cover show a stable, inverse relationship with heat island intensity, increasing steadily as heat island intensity decreases. However, some regulatory planning units in weak heat island areas still exhibit low green space and vegetation cover proportions. Conversely, strong heat island areas consistently have low levels of greenery, suggesting that low greening does not always result in high-intensity heat islands, but higher greening levels can mitigate heat island intensity.

The building cover proportion and impervious surface cover proportion are directly proportional to heat island intensity, decreasing steadily as heat island intensity decreases. In extremely strong heat island areas, the building and hardened surface cover proportions are as high as 47.14% and 30.28%, respectively, totaling 77.42%. In weak heat island areas, these proportions drop to 27.89% and 13.80%. This indicates that building cover and hardened surface cover significantly and consistently contribute to heat island intensity, making them important factors in inducing high-intensity heat islands.

3.4.3 Comparison of Building Intensity Indicators

The intensity of construction is reflected by the proportion of construction land, floor area ratio, and building density. Statistics show that the proportion of construction land and building density is directly proportional to the intensity of the heat island, demonstrating a significant regularity. Among them, the difference in building density across different heat island areas is relatively large, whereas the correlation between floor area ratio and heat island intensity is weak, with no significant differences in floor area ratio indicators among regulatory planning units in various heat island intensity areas.

Both the proportion of construction land and building density are directly proportional to heat island intensity. The numerical differences in these proportions across different heat island intensity areas are considerable and decrease as heat island intensity lessens. The average proportions of construction land in extremely strong, strong, relatively strong, and weak heat island areas are 77.94%, 69.50%, 62.21%, and 43.73%, respectively, while the average building densities are 32.85, 22.77, 17.63, and 11.72%. Notably, the stronger the heat island intensity, the smaller the change in the proportion of construction land. The difference in the proportion of construction land within regulatory planning units in weak heat island areas is significant, whereas it is very similar within extremely strong heat island areas. The variation in building density across different heat island intensity areas is relatively small and remains consistent, with fewer regulatory planning units in different heat island intensity areas exhibiting the same building density. This indicates that building density more significantly affects heat island intensity.

The low correlation between floor area ratio and heat island intensity is primarily due to a large number of spatial units with similar floor area ratios belonging to different heat island intensity areas, resulting in an unclear relationship between floor area ratio and heat island intensity. Statistics show that regulatory planning units with a floor area ratio between 0.6 and 1.3 are distributed across all four heat island intensity areas. However, regulatory planning units with extremely high or low floor area ratios exhibit significantly different heat island effects, predominantly found in extremely strong and weak heat island areas, respectively.

3.4.4 Comparison of Building Combination Indicators

The building combination of regulatory planning units is reflected by GSVF (Ground Surface Visible Fraction), roughness, and building shape coefficient. Statistics indicate that, compared to the aforementioned types of urban form indicators, the correlation between building combination indicators and heat island intensity is not strong, and none of these indicators exhibit a clear regular change with heat island intensity. The building shape coefficients of different heat island intensity areas are very similar, showing no discernible relationship with heat island intensity. While GSVF and roughness are roughly inversely and directly proportional to heat island intensity, respectively, there is no strong mapping relationship between their values and heat island intensity. Regulatory planning units with a GSVF between 0.88 and 0.96 and roughness between 0.11 and 0.25 are widely distributed across different heat island intensity areas. Interestingly, the extremely strong heat island area presents an anomaly, showing higher GSVF and lower roughness compared to the strong heat island area.

The weak correlation between GSVF, roughness, building shape coefficient, and heat island intensity suggests that these indicators may not have a linear relationship with heat island intensity. On one hand, this manifests as a limited influence and a vague mapping relationship. On the other hand, the GSVF and roughness in the extremely strong heat island area deviate from the general rule. This indicates that the formation of the heat island effect, particularly in extremely strong heat island areas, is strongly influenced by factors other than building combinations. Even when spatial openness and ventilation potential are conducive to reducing heat island intensity, many regulatory planning units still exhibit extremely strong heat island effects due to other influencing factors.

3.5 Core Urban Form Indicators Selection

3.5.1 Selection Principles

The "core urban form indicators" defined in this article refer to the urban form indicators that have a significant impact on the intensity of the heat island. To identify which indicators among the many alternatives are closely related to heat island intensity, the following three principles should be followed:

3.5.1.1 At Least One Indicator from Each Category of Urban Form Indicators Should be Selected

The literature review revealed that functional development, land surface coverage, construction intensity, and building combination are all urban form elements influencing the heat island effect. This indicates that the impact of these four types of urban form indicators on heat island intensity objectively exists. While not all urban form indicators within these categories impact heat island intensity, at least one indicator from each category should be included in the core urban form indicators that affect heat island intensity.

3.5.1.2 The Core Urban Form Indicators Should Change Regularly with the Change in the Intensity of the Heat Island

In different heat island intensity zones, as the intensity of the heat island increases, there is generally a monotonically increasing or decreasing trend. This trend can be used to identify which indicators change regularly with the heat island intensity. If an urban form indicator does not follow this principle, there are two possibilities: one is that the impact of this indicator on heat island intensity is minimal or non-existent, and therefore it does not show a regular correlation (such as floor area ratio); the second is that this indicator significantly influences heat island intensity only within a specific range of values, showing a regular positive or negative correlation. However, when it exceeds this range, no correlation occurs, or the correlation direction changes (from increasing to decreasing or vice versa). Urban form indicators that fall into these two categories either have little correlation with heat island intensity, or their correlation is too complex. Therefore, they are not included in the scope of the subsequent research.

3.5.1.3 Do not Contradict the Urban Form Characteristics of Typical Heat Island Areas

The urban form characteristics of the typical extremely strong heat island areas and typical weak heat island areas selected in the previous text are representative. The summary and comparison of these characteristics form the conclusions of qualitative research, carrying a certain degree of empiricism. After applying the second principle's screening, the remaining urban form indicators can be assumed to be related to the intensity of the heat island. If these correlation features are found to contradict the spatial morphology characteristics of the typical heat island areas, they need to be excluded.

3.5.1.4 Avoid Redundancy Between Different Indicators

In order to more comprehensively reveal the urban form indicators that affect the intensity of the heat island, the potential indicators set in this article inevitably include instances where different indicators redundantly describe the same performance attributes of the same spatial element. Including all these redundant indicators in subsequent research could complicate the model due to collinearity between the indicators during regression analysis. Therefore, the redundant indicators remaining after applying the aforementioned selection principles should be eliminated.

3.5.2 Selection Results

In the previous section, through a literature review and assessment of the functional characteristics of different heat island intensity areas, we pre-set four types of urban form elements that affect heat island intensity along with a series of related indicators. However, after comparing the urban forms in different typical heat island areas and various heat island intensity areas, it was found that some of these urban form indicators are more closely related to heat island intensity (Table 3.5):

Table 3.5 Core urban form indicators influencing urban heat island

Indicator type	Name of index	Data sources
Functional development indicators	The proportion of industrial land	Calculated from Wuhan city land status map
	The proportion of commercial land	Calculated from Wuhan city land status map
	population density	Calculated from Wuhan "Six Pu" data calculation
Surface cover class indicators	The proportion of covered soil	Calculated from Landsat8 remote sensing data
	NDVI	Calculated from Landsat8 remote sensing data
	The proportion of vegetation cover	Calculated from Landsat8 remote sensing data
	Ratio of green space	Calculated from Wuhan city land status map
Construction intensity category indicators	Building density	Calculated from current building vector data
	The proportion of construction land	Calculated from Wuhan city land status map
Building combination class index	GSVF	Calculated from current building vector data

3.5.2.1 Functional Development Indicators: Proportion of Industrial Land Use, Proportion of Commercial Land Use, Permanent Population Density

The three indicators in the "Functional Development" category—proportion of industrial land use, proportion of commercial land use, and permanent population density—are significantly related to heat island intensity, all showing a positive impact. On one hand, these indicators increase with the rise of heat island intensity across different heat island intensity areas. On the other hand, two of the three typical cases in the extremely strong heat island area primarily use land for industry and commerce. While the third case has a very high permanent population density. In contrast, none of the typical cases in the weak heat island area exhibit a high proportion of industrial land use, commercial land use, or permanent population density. Although the density of employment positions also shows regular changes across different heat island intensity areas, it contradicts the urban form characteristics of typical heat island intensity areas. Specifically, in some cases within the weak heat island area, the density of employment positions is high, while in some cases within the extremely strong heat island area, it is low. This indicates that the impact of employment position density on heat island intensity is not significant.

3.5.2.2 "Land Surface Cover" Indicators: NDVI, Proportion of Vegetation Cover, Green Space Rate, Proportion of Soil Cover

The indicators reflecting the ecological level in the "Land Surface Cover" category have a significant relationship with heat island intensity, while individual indicators reflecting impervious surfaces do not show a significant correlation with heat island intensity. Impervious surfaces include buildings, roads, and hardened surfaces. Although a comparison of urban forms across different heat island intensity areas found that the proportion of building cover and road cover has a certain positive relationship with heat island intensity, two of the three typical weak heat island area cases have extremely high proportions of building and road cover. Therefore, the proportion of building cover and road cover is excluded from consideration. Additionally, the impervious surface cover indicator overlaps with the NDVI indicator to some extent, so it is also excluded to avoid collinearity. The ecological level indicators—NDVI, proportion of vegetation cover, green space rate, and proportion of soil cover—show consistent positive or negative impacts on heat island intensity. These indicators regularly increase or decrease with changes in heat island intensity across different heat island intensity areas, and they exhibit significant differences between cases in the extremely strong and extremely weak heat island areas. Therefore, they are considered significant urban form indicators related to heat island intensity.

3.5.2.3 "Construction Intensity" Indicators: Building Density, Proportion of Construction Land Use

The indicators of "Construction Intensity," specifically building density and the proportion of construction land reflecting urban function abundance, demonstrate a significant positive correlation with heat island intensity. Conversely, the relationship between floor area ratio and heat island intensity lacks robustness. In comparing urban forms among typical heat island areas, notable disparities in building density emerge between typical extremely strong and weak heat island areas. Across different heat island intensity zones, the proportion of construction land in Fengsheng Road block, a typical extremely weak heat island area, is remarkably low, whereas, in typical extremely strong heat island areas, this proportion approaches 100% and increases with rising heat island intensity. The variability in floor area ratio and heat island intensity is evident in both typical extremely strong and weak heat island areas, characterized by instances of low floor area ratios; for example, both Zhuankou Industrial Zone and Fengsheng Road block exhibit low floor area ratios. Although floor area ratio generally correlates with heat island intensity in strong, stronger, and weak heat island areas, the notably low floor area ratios in extremely strong heat island areas do not conform to this pattern.

3.5.2.4 "Building Combination" Class Indicator: Ground Sky View Factor

Strictly speaking, the various indicators in the "Building Combination" category do not show as significant a relationship with heat island intensity as the aforementioned indicators. Among these, only the indicator of GSVF exhibits some correlation with heat island intensity. Statistical analysis reveals that a lower GSVF is associated with greater heat island intensity. However, this pattern does not hold in extremely strong heat island areas, where GSVF is notably high due to the large proportion of industrial areas. Despite this anomaly, a substantial body of literature establishes a correlation between GSVF, building combination methods, and heat island intensity. Therefore, GSVF is still included as a core urban form indicator affecting heat island intensity.

3.5.3 Credibility Test

Since some core urban form indicators do not conform to the normal distribution (such as the proportion of industrial land), the Pearson correlation coefficient cannot be used to test the relationship between core urban form indicators and heat island intensity. Instead, the Spearman correlation coefficient (Table 3.6) is employed. The statistical results show that all core urban form indicators passed the F-value test,

Table 3.6 Spearman correlation coefficient between core urban form indicators and urban heat island intensity

	Correlation coefficient
The proportion of industrial land	0.322***
The proportion of commercial land	0.256***
Population density	0.282***
The proportion of soil cover	−0.185***
NDVI	−0.454***
The proportion of vegetation cover	−0.419***
Ratio of green space	−0.318***
Building density	0.552***
The proportion of construction land	0.438***
GSVF	−0.076*

Note ***, **, and * indicate statistical significance at the 1%, 5%, and 10% levels, respectively

indicating that the screened core urban form indicators are significantly related to heat island intensity.

3.6 Summary

This chapter addresses a central question: Which urban form indicators have a significant impact on the heat island effect? First, this book synthesizes existing literature and conclusions from this chapter to propose four types of urban form indicators as potential influences on the heat island effect: functional development indicators, surface coverage indicators, construction intensity indicators, and building combination indicators, encompassing a total of 19 urban form indicators. The spatial differentiation of these indicators shows a certain coupling with the spatial pattern of the heat island, supporting the rationale for their selection.

Subsequently, the study performs a cluster analysis of various functional development indicators, selecting six blocks with distinct functional development characteristics from the research area as typical heat island zones. Detailed statistics and comparisons of their urban form indicators are conducted, revealing that typical strong heat island zones exhibit characteristics of "concentrated industrial, commercial, or high-density residential areas," "dominance of impermeable surfaces with limited ecological elements," "high building density," and "compact, closed, plate-like aggregation with low density and high sparsity" in building combinations. In contrast, typical weak heat island zones show characteristics of "sparse construction, single residential areas or mixed commercial-residential zones," "a mix of impermeable surfaces and green spaces with abundant ecological elements," "low building density," and "loose, open, point-like dispersion with uniform high and low density" in building combinations.

The study then compares the urban form indicators across different heat island intensity zones. By analyzing the heat island intensity from high to low—extremely strong heat island zones, strong heat island zones, relatively strong heat island zones, and weak heat island zones—the study summarizes the changing patterns of each urban form indicator. It is found that the proportions of industrial land, commercial land, permanent population density, job density, green space, NDVI, building coverage, road coverage, hardened surface coverage, soil coverage, vegetation coverage, construction land, and building density exhibit regular changes across these heat island intensity zones.

Finally, by integrating the urban form differences of typical heat island zones with the changing patterns of urban form indicators across different heat island intensities, and following relevant selection principles, ten urban form indicators are identified as having the most significant impact on heat island intensity. These indicators are: proportion of industrial land, proportion of commercial land, permanent population density, proportion of soil coverage, NDVI, proportion of vegetation coverage, green space rate, building density, proportion of construction land, and GSVF. The reliability of these indicators is then tested.

References

1. Pin W (2015) Study on thermal environment optimization of urban central business district in hot and humid area. South China University of Technology
2. Chun B, Guldmann JM (2014) Spatial statistical analysis and simulation of the urban heat island in high-density central cities. Landsc Urban Plan 125:76–88
3. Wenjing L, Yanuar S, Perry PJ (2004) GIS analysis for the climatic evaluation of 3d urban geometry: the development of GIS analytical tools for sky view factor

Chapter 4
The Mechanism of Urban Form Affecting Heat Island Effect

This chapter employs a combination of qualitative and quantitative methods to elucidate the mechanisms through which urban form indicators affect heat island intensity. The chapter begins by constructing an explanatory framework for these impact mechanisms, proposing that the influence of urban form on heat island intensity can be categorized into intensification mechanisms, mitigation mechanisms, and external environmental modification mechanisms. Subsequently, it analyzes each mechanism in detail by combining the quantitative correlation rules between urban form indicators and heat island intensity. This analysis identifies three intensification mechanisms and two mitigation mechanisms, along with their associated spatial effects. Finally, the chapter proposes a comprehensive impact mechanism that integrates these intensification and mitigation effects while considering external environmental factors. By analyzing the correlation between the spatial effects of this integrated mechanism and the actual observed values of heat island intensity, the chapter verifies the rationality of the comprehensive impact mechanism and compares the contributions of different urban form indicators to heat island intensity.

It should be noted that while the four urban form elements—functional development, surface coverage, construction intensity, and building combination—are all closely related to heat island intensity, only some indicators show a significant relationship with heat island intensity. Therefore, the exploration of mechanisms will focus on the core urban form indicators identified in this chapter.

Y. Lu, *Urban Form and Urban Heat Island*, Urban Sustainability,
https://doi.org/10.1007/978-981-96-3541-2_4

4.1 A Framework of the Mechanism Explanation

4.1.1 Theoretical Basis: Energy Balance in Local Canopy Space

Scholars' discussions on the causes of heat islands primarily focus on two scales: the Urban Canopy Layer (UCL) and the Urban Boundary Layer (UBL) [1]. The UCL, often referred to as the building layer, encompasses the range from the ground up to approximately the average height of rooftops. This layer is closely related to building height, density, geometric shape, and building materials, as well as the extent of green areas. The UBL, on the other hand, is a medium-scale atmospheric layer affected by the Earth's surface friction, thermal processes, and evaporation, extending from the building rooftops to the middle height of the cumulus clouds. It is influenced by the thermal and dynamic effects of uneven roofs, the quality of urban air, and exchanges of materials and energy with the UCL, as well as by the surrounding environment. The focus of the UCL is on human factors and the direct impact of buildings and activities, while the UBL's focus is more macroscopic, with its thermal environment shaped by the heterogeneity of the underlying surface, including land-sea distribution, oases, and terrain undulations. Given the research objectives of this book, the discussion on the mechanisms of heat island formation will be confined to the UCL.

The impact of the urban canopy on the local thermal environment primarily involves the effects of the urban underlying surface and human heat emissions. The urban underlying surface, composed of hard, dense, dry, and impermeable materials such as bricks, asphalt, and metal plates, converts radiant energy into thermal energy. Additionally, the three-dimensional structure of building clusters affects solar shortwave radiation, ground longwave radiation, and air turbulence within the canopy, thereby influencing heat island intensity. Human heat emissions, including pollutants and greenhouse gases from fossil fuel combustion, as well as heat from industrial activities and air conditioning, further contribute to the heat island effect (Table 4.1).

From the perspective of energy balance, the urban area can be viewed as a spatially closed system. The fundamental cause of heat island formation is the imbalance between heat absorption and heat dissipation, which leads to the accumulation of heat (Fig. 4.1). In the dynamic energy balance, the increased "income" of energy relative to the decreased "expenditure" results in higher heat island intensity. The energy balance diagram in Table 7 illustrates that during the formation of a heat island, the local canopy's capacity to absorb energy increases while its ability to dissipate energy decreases, thereby increasing the heat island intensity.

Table 4.1 Causes of urban heat island in urban canopy

Energy absorption and distribution situation	Energy revenue and expenditure path	Cause
Increase energy absorption	Increase the absorption of short-wave radiation	Building combination increases surface area and repeated refraction
	Increase the long-wave radiation from the sky	Air pollutants increase the absorption and reemission of radiation
	Anthropogenic heat emission	Buildings and roads increase combustion emissions
	Increase heat storage	Building materials increase heat storage
Inhibit the dissipation of energy	Reduce the long-wave radiation loss	Building combination reduces sky visibility
	Reduce evaporation and heat dissipation	Building materials reduce water evaporation
	Reduce the total turbulent heat transport	Closed building combinations reduce air convection

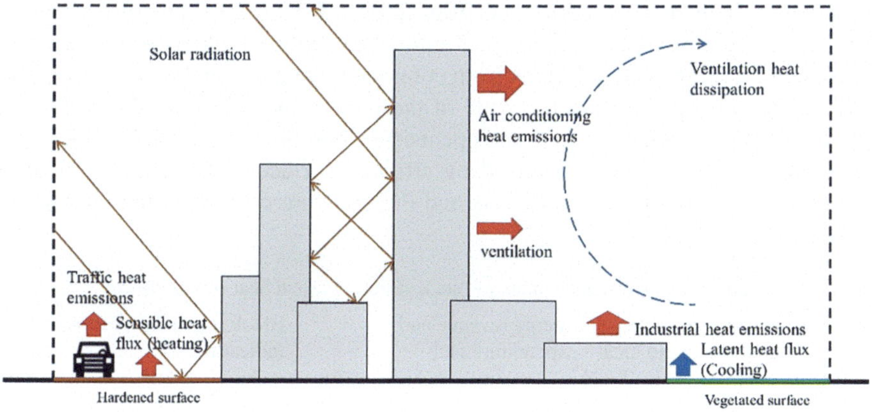

Fig. 4.1 Diagram of energy budget in local urban canopy

4.1.2 *Heat Conduction Mode: Human Heat Emission, Radiant Heat Gain, Convective Heat Dissipation*

According to the basic laws of thermodynamics, both heat accumulation and dissipation processes fall within the category of heat transfer, which always involves three fundamental modes: heat conduction, convective heat transfer, and radiant heat.

For an object, there are three positional relationships with other objects: (1) within the object, (2) objects in direct contact with it, and (3) objects at a distance from it.

As long as there is a temperature difference, heat conduction will occur in these three relationships. Generally, heat transfer within solids or between solids in contact is referred to as "heat conduction," while heat transfer in fluids or between solids and fluids is classified as "convective heat transfer." Heat transfer between two objects without physical contact or a medium is termed "radiant heat." In the urban canopy space, heat conduction can be ignored in the energy balance, leaving convective heat transfer and radiant heat as the primary modes of energy transfer. According to Newton's cooling law, the rate of convective heat transfer is directly proportional to both the convective heat transfer area and the temperature difference, while radiant heat is significantly influenced by the relative positions of objects.

In the urban canopy space, anthropogenic heat emission, evaporative cooling, and turbulent heat transport are classified under "convective heat transfer," whereas absorption and refraction of shortwave radiation, longwave radiation, and sensible heat storage fall under "radiant heat" (Table 4.1). Among these, anthropogenic heat emission is categorized as a basic mechanism of energy balance, while evaporative cooling and turbulent heat transport are grouped under "convective cooling," and all forms of radiant heat transfer are classified as "radiant heat" (Table 4.2).

To determine how block thermal performance indicators influence these mechanisms in the urban canopy space (Table 4.2), it is necessary to assess several factors. First, the function type of a building affects the types of anthropogenic heat emissions, which include ventilation heat from industrial buildings, power heat emissions, chimney emissions, radiator heat emissions, cooling water heat emissions, and air conditioning heat emissions from commercial and residential buildings. The building area also determines the scale of these emissions. Second, the building's light-receiving area, shadow area, and openness of building arrangements determine the extent of solar radiation received by artificial surfaces. Third, the ventilation performance of building arrangements and the moisture content of the underlying

Table 4.2 Urban form performance index of block affecting urban heat island effect

The basic mechanism of the energy expenditure in the local canopy space	Energy revenue and expenditure path	Block thermal performance indicators
Artificial heat removal mechanism	Anthropogenic heat emission	Functional type of building Floor area
Radiation heat gain mechanism	Increased absorption of shortwave radiation Increased long-wave radiation from the sky Increase sensible heat storage Reduce long-wave radiation loss	Thermal properties of artificial materials Building surface area Building shadow area The openness of the building mix
Convection heat dissipation mechanism	Reduced evaporative heat dissipation Reduce total turbulent heat transport	Ventilation performance of the building combination Underlying surface moisture content

surface affect the scale of turbulent heat transport and evaporative cooling, which in turn dissipate heat through convective cooling mechanisms.

4.1.3 Mechanism Framework: Enhancement Mechanism, Damping Mechanism, and External Environment Factor Correction Mechanism

4.1.3.1 Mechanism Framework Structure

The three basic heat transfer methods in the local canopy space—namely, the artificial heat dissipation mechanism, convection heat dissipation mechanism, and radiation heat absorption mechanism—affect both heat absorption and heat dissipation, thereby determining the overall energy balance. In the "energy balance path," artificial heat emission, increased absorption of short-wave radiation, and increased sensible heat storage contribute to heat absorption, while increased evaporation heat dissipation, reduced sensible heat storage, increased long-wave radiation loss, and increased turbulent heat transfer enhance heat dissipation. This book categorizes these processes into "mechanisms of urban form increasing heat island intensity" (referred to as "intensification mechanisms") and "mechanisms of urban form reducing heat island intensity" (referred to as "mitigation mechanisms").

Additionally, the heat island intensity of each management unit is influenced by external environmental factors. Broadly, the external environment encompasses all elements affecting heat island intensity, such as atmospheric conditions and pollutants. However, this book focuses on external urban form patterns, such as water systems, roads, and concentrated industrial areas, rather than the external environment in a broad sense. Analyzing the impact of these external factors on heat island intensity is crucial for refining research conclusions about the correlation between urban form and heat island intensity. This involves introducing a correction mechanism based on intensification and mitigation mechanisms, with core urban form indicators as explanatory variables and external environmental factors as control variables, thereby forming an explanatory model for heat island intensity (Fig. 4.2a).

Combining quantitative analysis, the ten urban form indicators are divided into two categories: indicators that increase heat island intensity and indicators that reduce heat island intensity (Fig. 4.2b). Notably, GSVF is unique in that it has both intensifying and mitigating effects on heat island intensity. Each indicator affects heat island intensity through three heat transfer mechanisms: artificial heat dissipation, radiation heat absorption, and convection heat dissipation.

First, functional development indicators relate to building function, building area, and the thermal properties of materials, impacting heat island intensity through artificial heat dissipation and radiation heat absorption mechanisms. These indicators are directly proportional to heat island intensity, as buildings with industrial and commercial functions generate air conditioning heat dissipation, ventilation, power

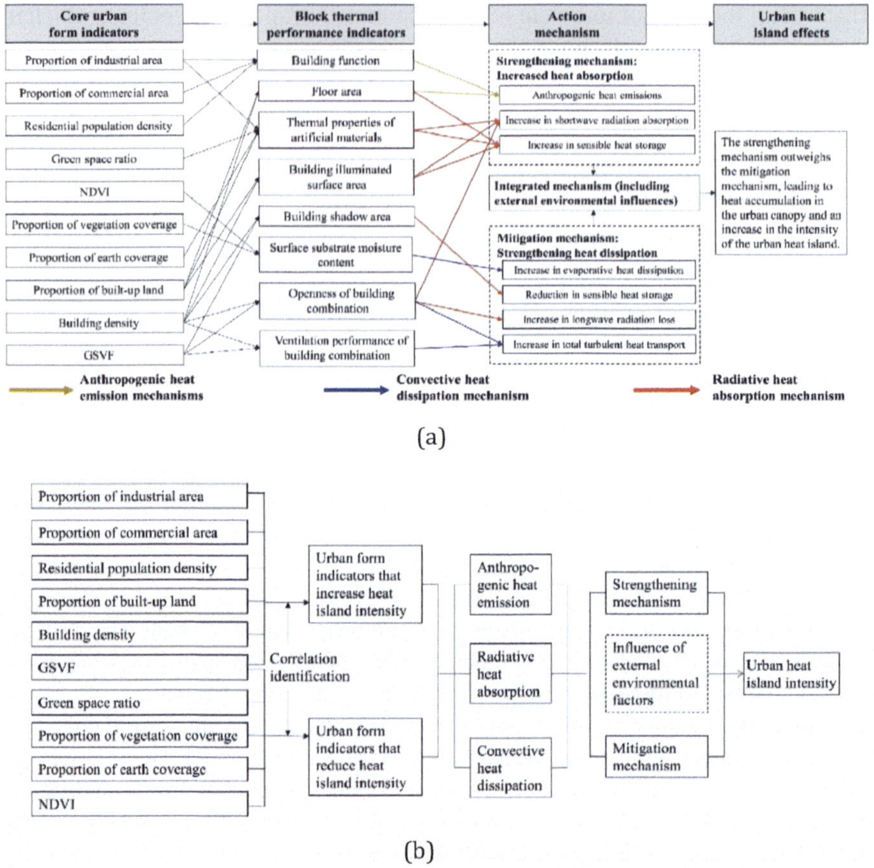

Fig. 4.2 Impact of core urban form indicators on urban heat island intensity

heat dissipation, chimney heat dissipation, radiator heat dissipation, and cooling water heat dissipation, affecting local canopy space heat absorption. Additionally, higher permanent population density increases residential activity heat dissipation. Industrial areas with more metal sheets, which have low reflectivity, absorb short-wave radiation effectively, whereas high green space rates reduce radiation heat absorption.

Second, land surface cover indicators relate to the thermal properties of materials and surface moisture content, influencing heat island intensity through the convection heat dissipation mechanism. These indicators are inversely proportional to heat island intensity, as ecological land surfaces with high moisture content absorb heat through continuous evaporation, mitigating heat island intensity.

Third, Construction intensity indicators are associated with building light-receiving area and floor area, affecting heat island intensity through artificial heat dissipation, radiation heat gain, and convection heat dissipation mechanisms. These indicators are directly proportional to heat island intensity, with larger light-receiving areas increasing energy absorption and building areas enhancing sensible heat storage.

Fourth, building combination indicators relate to shadow area and the openness of the building arrangement (reflected by GSVF), affecting heat island intensity through radiation heat gain and convection heat dissipation mechanisms. Lower openness reduces long-wave radiation decay, inhibiting radiation heat gain, while better building combination permeability facilitates internal heat dissipation. Ventilation performance is generally inversely proportional to GSVF.

Finally, areas with extreme land surface temperatures are analyzed as external environmental factors to correct the correlation between urban form indicators and heat island intensity, assuming these factors influence the surrounding thermal environment. Based on the fundamental heat conduction mechanisms, convection heat dissipation allows for mutual heat transfer between adjacent areas with temperature differences.

4.1.3.2 Correlation Between Urban Form Indicators and Urban Heat Island Intensity

This book identifies the correlation between core urban form indicators and heat island intensity from both "single form indicator" and "multiple form indicators" perspectives. Simple linear regression is employed to analyze the correlation between each single urban form indicator and heat island intensity under the "intensification mechanism" and "mitigation mechanism," reflecting each indicator's effect on heat island intensity in isolation. The "comprehensive mechanism" uses multiple linear regression to assess the effect of each indicator on heat island intensity while considering external environmental factors.

Initially, 589 regulatory planning units were sampled, and we applied simple linear regression to identify the correlation between each urban form indicator and heat island intensity. This method reveals each single urban form indicator's impact on heat island intensity. Although Chapter 4 shows the global correlation direction, it does not confirm significant correlations in different value ranges. To enhance regression analysis objectivity, indicator segmentation analysis examines changes in the correlation between urban form indicators and heat island intensity across value ranges. Simple linear regression is performed within each management unit category to analyze whether each urban form indicator significantly correlates with heat island intensity in different value ranges and calculate correlation coefficients. Furthermore, given that urban form impacts heat island intensity through multiple factors, multiple linear regression is used to include all core urban form indicators and external environmental factors in the model, comparing the effects of each indicator on heat island intensity. Finally, to identify external environmental factors affecting heat island intensity, the study combines linear regression results of urban form indicators with actual heat island intensity observations to obtain simulated and observed heat island intensity values.

4.2 Mechanism of Urban Form Increasing Heat Island Intensity

4.2.1 Function Operation Increases Heat and Temperature: Industrial, Commercial, Residential Function Operation Increases Man-Made Heat Discharge

4.2.1.1 Mode of Urban Form Action on the Thermal Environment

Industry, commerce, and residence are the primary sources of urban man-made heat discharge. A study of heat emissions in 12 cities in the Yangtze River Delta reveals that industrial enterprises contribute the highest amount of man-made heat emissions, accounting for 75.1% of the total, while commercial and residential buildings together account for 9.9% of the total. The contribution of human metabolism is minimal in comparison and can be considered negligible. Therefore, the heat emissions resulting from industrial functions (represented by the proportion of industrial land), commercial functions (represented by the proportion of commercial land), and residential functions (represented by the permanent population) are key components of urban man-made heat discharge.

The mechanisms through which these functions increase heat island intensity are as follows:

- Industrial Land. Industrial processes, such as those involving melting furnaces and boilers, generate significant amounts of waste heat, which is discharged into the environment via chimneys, radiators, and cooling water systems. Consequently, the proportion of industrial land is positively correlated with chimney heat discharge, radiator heat discharge, and cooling water heat discharge. Additionally, industrial areas frequently use metal sheets, which have high short-wave radiation absorption rates compared to the brick and stone materials commonly used in other functional land types.
- Commercial Land. Commercial land encompasses functions like office spaces, retail, and wholesale activities, which result in substantial air conditioning heat discharge and extensive indoor and outdoor ventilation. While air conditioning primarily regulates indoor temperatures, it contributes to urban canopy heating, thus exacerbating the urban heat island effect. For instance, in the summer of 2001, Beijing's air conditioning power consumption reached 2 million kW, equivalent to the heat from burning 800,000 kg/h of coal, all of which was released into the urban canopy, thus intensifying the heat island effect.

- Permanent Population Density. The density of the permanent population reflects residential activity intensity, with heat emissions arising from household energy consumption for cooking, appliances, heating, and cooling (Table 4.3). This contributes to the heat island effect through both air conditioning heat dissipation and ventilation. Urban residents consume more energy compared to rural residents due to lifestyle differences. The average total household energy consumption in urban areas is 1506.85 kgce/year, with a per capita consumption of 653.00 kgce/year, whereas in rural areas, the average total energy consumption is 1100.41 kgce/year, with a per capita consumption of 446.69 kgce/year. Urban household energy consumption is approximately 1.4 times that of rural households.

Table 4.3 Standard China household energy consumption balance table (2012, unit: kg of standard coal)

Energy activities	Energy type	Total energy consumption
Cooking	Kitchen range	294.71
	Electric cooker	21.12
	Microwave oven	4.38
	Coal-scuttle	0.24
	Bread machine	0.18
	Autoclave	4.59
	Other	1.78
Domestic appliance	Refrigerator	21.84
	Refrigerator	1.33
	Washing machine	10.01
	Goggle-box	28.34
	IBM PC	20.65
	Lighting bulb	20.11
Home heating	Central heating	647.02
	Independent heating-air conditioning	3.95
	Independent heating-electric heating/oil heating system	6.53
	Independent heating-boiler piping	17.71
	Independent heating-furnace	88.47
	Water heater	204.74
Home refrigeration	Electric fan	2.26
	Air-conditioning	25.94
Total energy consumption		1425.92

4.2.1.2 Degree of Impact

On a global scale, the results of the simple linear regression indicate that the proportion of industrial land, commercial land, and the density of the permanent population all have a significant linear relationship with heat island intensity, with a direct proportionality (Table 4.4). However, segmented regression analysis reveals that these indicators do not exhibit a significant positive correlation with heat island intensity across all numerical ranges (Table 4.5): ① In regulatory planning units where the proportion of industrial land exceeds 40%, it significantly intensifies heat island intensity, with an increase of 0.41 °C for every 10% increase in the proportion of industrial land; ② When the proportion of commercial land is less than 5% or greater than 15%, its impact on heat island intensity is significant but shows positive and negative correlations, respectively. Further analysis indicates that in areas with a smaller proportion of commercial land, business types such as retail and entertainment generate substantial human heat emissions, is primarily for hotels with higher levels of greening; ③ The density of the permanent population has a significant intensifying effect on heat island intensity in the range of 1.0 to 1.5 thousand people/km^2, with an increase of 0.26 °C for every 0.1 thousand people/km^2 increase in population density. Although there is still a slight intensifying effect at densities greater than 2.5 thousand people/km^2, the increase is only 0.03 °C per 0.1 thousand people/km^2.

This analysis shows that while each functional development index has a significant positive effect on heat island intensity, the extent of this effect is conditional and concentrated within specific ranges of the index values. For industrial land, the proportion of industrial land only significantly affects heat island intensity when it exceeds 40%, due to the limited ventilation capacity that prevents the effective dissipation of waste heat produced by industrial activities. In regulatory planning units where industrial land constitutes less than 40%, the waste heat can be dissipated effectively, and the proportion of industrial land is not a major factor in the heat island effect. Similarly, although there is a significant positive correlation between commercial land and heat island intensity on a global scale, this effect is primarily observed when the proportion of commercial land is less than 5%. The effect of permanent population density on heat island intensity is significant mainly in the range of 1.0 to 1.5 thousand people/km^2; at higher densities, the effect diminishes, with a very small regression coefficient of 0.03 °C per 0.1 thousand people/km^2 for densities greater than 2.5 thousand people/km^2. This suggests that in areas with low

Table 4.4 Linear regression coefficient of each core urban form indicators and urban heat island intensity based on 589 regulation planning units

Core urban form indicators	Linear regression coeffficient
The proportion of industrial land (%)	3.796***
The proportion of commercial land (%)	3.792***
Population density (ten thousand people/Km^2)	0.227***

Note ***, **, * means passing 1, 5 and 10% significance level test respectively

Table 4.5 Linear regression coefficients of urban form indicators and urban heat island intensity in different value intervals

Indicator name and unit of measurement	The value range of the indicator	Linear regression coeffficient
The proportion of industrial land (%)	Less than 5	5.755
	5–10	−3.753
	10–20	−1.925
	20–30	6.822
	30–40	−11.653
	More than 40	4.081**
The proportion of commercial land (%)	Less than 5	7.725*
Population density (ten thousand people/km^2)	Less than 0.5	0.540
	0.5–1.0	−0.553
	1.0–1.5	2.600*
	1.5–2.0	0.791
	2.0–2.5	0.502
	More than 2.5	0.299**

Note ***, **, and * indicate statistical significance at the 1%, 5%, and 10% levels, respectively

permanent population density, better ventilation allows for heat dissipation. In areas with higher population density, the heat island intensity is influenced more by other spatial factors.

4.2.2 *Artificial Surface Heat Exchange Warming: Buildings, Impervious Surfaces Absorb Solar Radiation, Increasing Radiative Heating*

4.2.2.1 Mode of Action

The thermal properties of artificial impervious materials differ significantly from those of ecological surfaces, leading to stronger radiative heating capabilities for the former. When evaluating the temperature rise of a material under solar radiation, two main factors are considered: the heat loss during the radiative heating process and the extent of temperature increase for a given amount of absorbed heat. Heat loss during radiative heating includes emission radiation loss and reflection radiation loss. Artificial impervious materials have lower emissivity compared to ecological surfaces, resulting in less emission radiation loss. Additionally, the reflectivity of artificial hardening materials is nearly identical to that of ecological surfaces (Table 4.6), which means that under the same solar radiation conditions, artificial hardening

Table 4.6 Albedo and emissivity of different surface covering materials

Surface	Pitch	Cement	Fragment of a brick	Forest	Grass	Wet soil	Dry soil
Albedo	0.05–0.20	0.10–0.35	0.20–0.40	0.07–0.20	0.15–0.30	0.10–0.25	0.20–0.40
Emissivity	0.95	0.71–0.90	0.13–0.28	0.98	0.96	0.98	0.90–0.95

Source [Israel] Ivy Erell, David Perlmut, [Australia] Terry Williamson, 2014

materials absorb more heat than ecological surfaces. The extent of temperature rise for a material, given the same amount of absorbed heat, depends on its specific heat capacity. Due to the high water content and specific heat capacity of forests, grasslands, and soils, while artificial materials have lower specific heat capacities, artificial hardening materials will generally experience a greater temperature increase and thus create a stronger heat island effect. In summary, under identical solar radiation conditions, artificial hardening materials, due to higher heat absorption and lower heat requirements for temperature increase, will have higher temperatures compared to ecological surfaces.

According to the association path between each construction intensity index and heat island intensity, the mechanisms promoting heat island intensity are as follows:

- Proportion of Construction Land: Construction land, as a concentrated area of urban activity with high energy consumption, shows a significant positive correlation with heat island intensity. The high proportion of construction land indicates extensive buildings and hardened surfaces that absorb and store more solar radiation due to a larger light-exposed area.
- Building Density: Higher building density increases the proportion of roof area (Fig. 4.3), and roofs are typically constructed from high thermal conductivity materials such as brick, cement, and reinforced concrete. These materials heat up quickly under solar radiation, increasing the rate of heat radiation within the block. Existing research shows that as building density and height-to-width ratio increase, reflectivity decreases, and the total heat gain of the building combination rises (Xin 2015). Additionally, higher building density reduces the space for permeable ground, limiting surface evaporation and thereby hindering convective heat dissipation.

4.2.2.2 Degree of Influence

The results of global-scale univariate linear regression indicate that both building density and the proportion of construction land have a significant positive linear relationship with heat island intensity. Notably, the impact of building density on heat island intensity is substantially higher than that of construction land (Table 4.7). However, segmented regression results reveal that these indicators do not uniformly correlate with heat island intensity across all value ranges (Table 4.8).

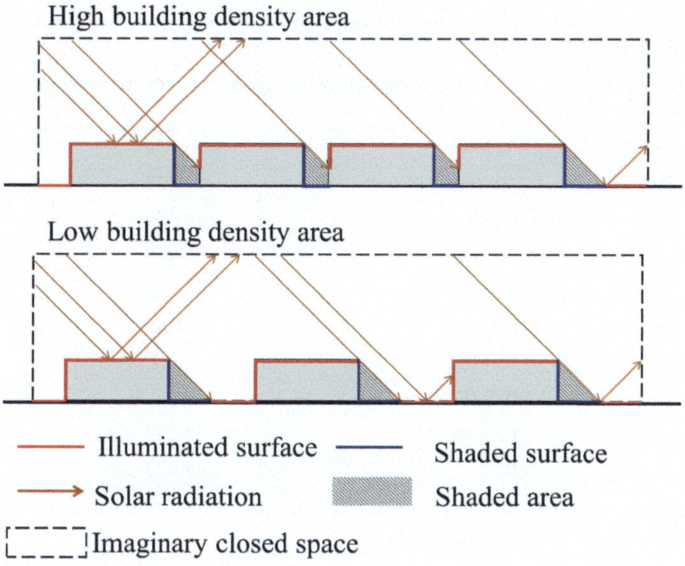

Fig. 4.3 Impact of solar radiation on building combinations of different building densities

Table 4.7 Linear regression coefficient of urban form indicators and urban heat island intensity of each core based on 589 regulatory planning units

Core urban form indicators	Linear regression coefficient
Building density	7.177***
The proportion of construction land (%)	2.761***

Note ***, **, * means passing 1, 5 and 10% significance level test respectively

The proportion of construction land does not show a significant correlation with heat island intensity across different value ranges.

The impact of building density on heat island intensity is primarily concentrated in areas with densities less than 0.1 and greater than 0.25, both of which exhibit a strengthening effect on heat island intensity. The most significant impact occurs in areas where building density is less than 0.1, where an increase of 0.1 in building density results in a 1.19 °C rise in heat island intensity. In areas with building densities between 0.25 and 0.35, an increase of 0.1 in building density corresponds to a 0.88 °C rise in heat island intensity.

These findings indicate that while both building density and the proportion of construction land generally enhance heat island intensity on a global scale, the strengthening effect of building density is concentrated in specific value ranges. In contrast, the correlation between the proportion of construction land and heat island intensity is not significant across all value ranges, suggesting that its positive correlation with heat island intensity is only a general trend on a global scale.

Table 4.8 Linear regression coefficients of urban form index and urban heat island intensity in different value intervals

Indicator name and unit of measurement	The value range of the indicator	Linear regression coeffficient
The proportion of construction land (%)	Below 40	0.754
	40–60	2.872
	60–70	5.667
	70–85	1.436
	Above 85	−2.460
Building density	0.1	11.856***
	0.1–0.2	−0.158
	0.2–0.25	−1.476
	0.25–0.35	8.753***
	0.35	4.672*

Note In parentheses are non-standard regression coefficients; *Note* ***, **, * means passing 1, 5 and 10% significance level test respectively

For building density, when it is less than 0.1, areas are predominantly composed of ecological permeable surfaces with generally low heat island intensity. In such areas, the few buildings present receive solar radiation and increase their surface temperature, becoming the primary heat source, thereby exhibiting a significant positive linear relationship. In the range of 0.1 to 0.25, the increase in building density results in a greater solar radiation-receiving interface, but this effect is counterbalanced by the increased shadow area, leading to no significant linear relationship within this range.

4.2.3 Building Combination Heat Concentration: Building Combination Slows Air Circulation, Restricting Convective Heat Dissipation

4.2.3.1 Mode of Action

Building combinations are a major component of urban local ventilation corridors, affecting the intensity of heat islands through convective heat dissipation mechanisms. To demonstrate how differences in the GSVF index affect the ventilation effect of building combinations, this book uses several typical cases from Chapter 4 as research subjects and employs Phonics software to simulate the wind environment of different typical heat island areas (Figs. 4.4, 4.5, 4.6, 4.7, 4.8 and 4.9).

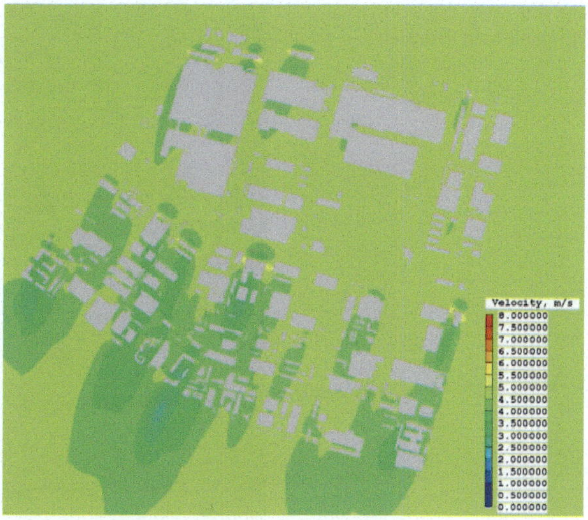

Fig. 4.4 Wind speed map of Zhuankou industrial park

Fig. 4.5 Wind speed map of Hanxi market

Fig. 4.6 Wind speed map of Minquan road district

Fig. 4.7 Wind speed map of Fengsheng road district

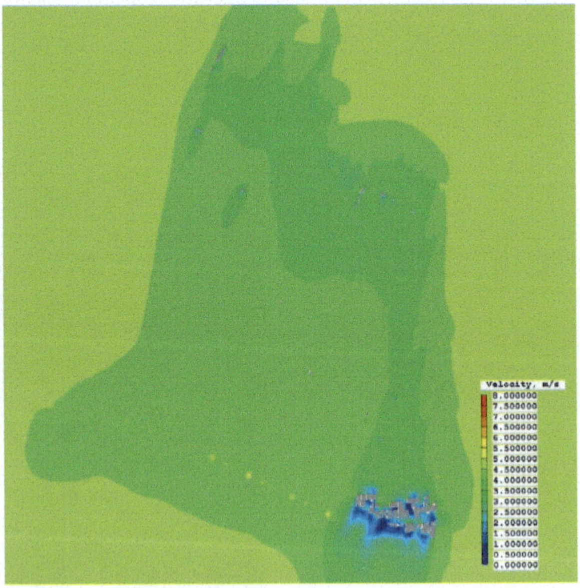

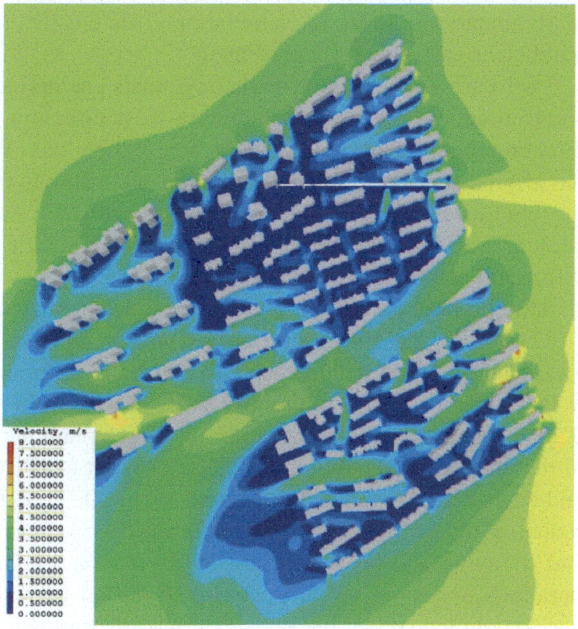

Fig. 4.8 Wind speed map of Xingye road district

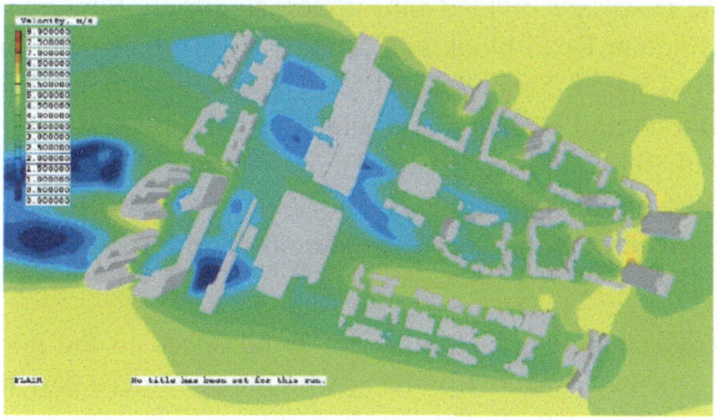

Fig. 4.9 Wind speed map of Wuhan Tiandi district

The factors involved in this model are as follows. First, Environmental Boundary Conditions: Geographical environment, wind environment, and thermal environment. Second, building Space Combination: Three-dimensional vector data of buildings. Third, material Properties: Various textures and thermal property parameters of building skin materials and underlay materials.

Environmental boundary conditions encompass model range, wind direction, wind speed, solar angle, solar radiation (W/m^2), and material property parameters. The scale of the simulated wind environment affects the simulation results. To ensure comparability, the book first calculates the projection lengths of each case in the east–west, north–south, and vertical directions, and then multiplies each by 4.0, setting the boundary range for the simulation. This ensures the tail flow of the simulated wind environment is fully presented.

Each case's grid division method is consistent, divided into 700 units in the north–south direction, 600 units in the east–west direction, and 20 units in the vertical direction, with further subdivisions within the block. We assume each case lay on a flat ground surface, with wind direction according to roughness calculation. We set the wind speed at 10 m from the ground as 4 m/s. The solar angle is the instantaneous angle at 10:55 on October 6, 2014, with direct solar radiation set at 500 W/m^2 and diffuse reflection at 250 W/m^2.

The building space combination is modeled in SketchUp software based on Wuhan city building data. The ground surface emissivity of the underlay is calculated based on the NDVI value within each case, with the solar absorption index uniformly set to 0.65. The building's emissivity and solar absorption index are set to 0.9862 and 0.72, respectively, referring to relevant design planning. The heat transfer coefficient of the underlay and buildings in all cases is set to 10 W/m^2/°C, with initial material and environmental temperatures both set to 20 °C. After setting the conditions, the radiation model is loaded, the spatial grid is refined according to the case scale, and the wind and heat environment simulation results are obtained through 1000 iterations to determine the spatial distribution of wind speed (m/s) on the 3.0 m horizontal plane under the same external environment in each case.

According to the wind speed simulation results for each typical heat island area, the cases are divided into three categories. First, significant Wind Speed Decay: Minquan Road block and Xingye Road block, with many areas approaching zero wind speed. Second, localized Calm Wind Phenomena: Hanxi Building Materials Market and Wuhan Tiandi block, usually downwind of high-rise buildings or areas with high building density. Third, Minimal Wind Speed Decay: Zhuankou Industrial Zone and Fengsheng Road block, show good ventilation potential.

Specific wind speed simulation results for each typical heat island area are as follows:

- Zhuankou Industrial Zone: Wind speed in the northern part is consistent with the external environment, while the southern part shows a slight decrease due to building orientation and height.
- Hanxi Building Materials Market: Similar to Zhuankou, with low-rise buildings in the north and medium/high-rise buildings in the south causing wind speed decreases and localized calm wind areas.
- Minquan Road Block: High building density leads to significant wind speed reduction in public spaces, but the street direction aligning with the main wind direction minimizes the calm wind range.
- Fengsheng Road Block: Low-rise buildings in the southeast cause wind speed reductions, but the overall area maintains wind speed consistency due to surface roughness.
- Xingye Road Block: The most significant wind speed decay among all areas, with large calm wind areas in densely built regions.

- Wuhan Tiandi Block: Staggered building heights with spacious ventilation channels maintain wind speed, preventing significant decreases.

The simulation results show that the area of calm wind regions in each typical heat island area is related to the GSVF index. Areas with lower GSVF exhibit compact, closed, plate-like aggregated building features, poor block ventilation potential, and generally large areas of calm wind.

4.2.3.2 Degree of Effect

The results of linear regression indicate that the linear correlation between GSVF and heat island intensity is weak, showing no significant correlation on a global scale. A significant correlation is only observed in areas where GSVF values are greater than 0.95 (Table 4.9). This suggests that the mechanism by which GSVF impacts heat island intensity is complex and cannot be solely attributed to reduced ventilation restricting heat dissipation. The statistical results do not support a significant positive or negative correlation, underscoring the multifaceted nature of this relationship.

4.2.4 Different Urban Forms Increase the Effect of Heat Island Intensity

The above three types of effects have all intensified the heat load in local areas. To analyze the mechanism of each urban form index on heat island intensity, this book scores each management unit based on the values of different urban form indices (Table 4.10). This scoring system reflects the spatial effect of each index on heat island intensity. The indices considered include the proportion of industrial land, the proportion of commercial land, the density of permanent residents, the proportion of construction land, and building density, all of which have a positive

Table 4.9 One-dimensional linear regression results of ground surface view factor index and heat island intensity in different value intervals

Indicator name and unit of measurement	The value range of the indicator	Linear regression coeffficient
GSVF	Less than 0.8	5.477
	0.8–0.85	11.373
	0.85–0.9	−3.677
	0.9–0.95	4.428
	More than 0.95	-18.614^{**}

Note In parentheses are non-standard regression coefficients; *Note* ***, **, * means passing 1, 5 and 10% significance level test respectively

effect on heat load. Consequently, higher values for these indices correspond to higher scores. Conversely, areas with lower GSVF values experience reduced ventilation, exacerbating the heat load, so higher GSVF values result in lower scores (Table 4.10).

By aggregating the scores of the indicators that enhance the heat load, a spatial effect map of the urban form index strengthening the heat island effect is created. As illustrated in Fig. 4.10, the spatial distribution of the strengthening mechanism scores closely resembles the intensity distribution of the heat island. The highest scoring areas (13.16 points) are concentrated within the first ring road and gradually decrease as the distance from the ring road increases, with protrusions in peripheral industrial areas. This indicates that the strengthening effect of urban form on heat island intensity is highly concentrated in the urban core and industrial zones, whereas areas with less construction land exhibit a smaller strengthening effect.

Table 4.10 The scores of urban form indicators that enhance urban heat island intensity

Urban form indicators	Value interval	Valuation
The proportion of industrial land	More than 50%	4
	25–50%	3
	10–25%	2
	0–10%	1
	0%	0
The proportion of commercial land	15%	3
	5–15%	2
	0–5%	1
	0%	0
Population density	More than 5.0	4
	2.5–5.0	3
	1.0–2.5	2
	0.5–1.0	1
	Less than 0.5	0
Building density	More than 75%	3
	60–75%	2
	30–60%	1
	Less than 30%	0
The proportion of construction land	More than 75%	3
	60–75%	2
	30–60%	1
	Less than 30%	0
GSVF (reduced convection heat dissipation)	Less than 0.95	1
	More than 0.95	0

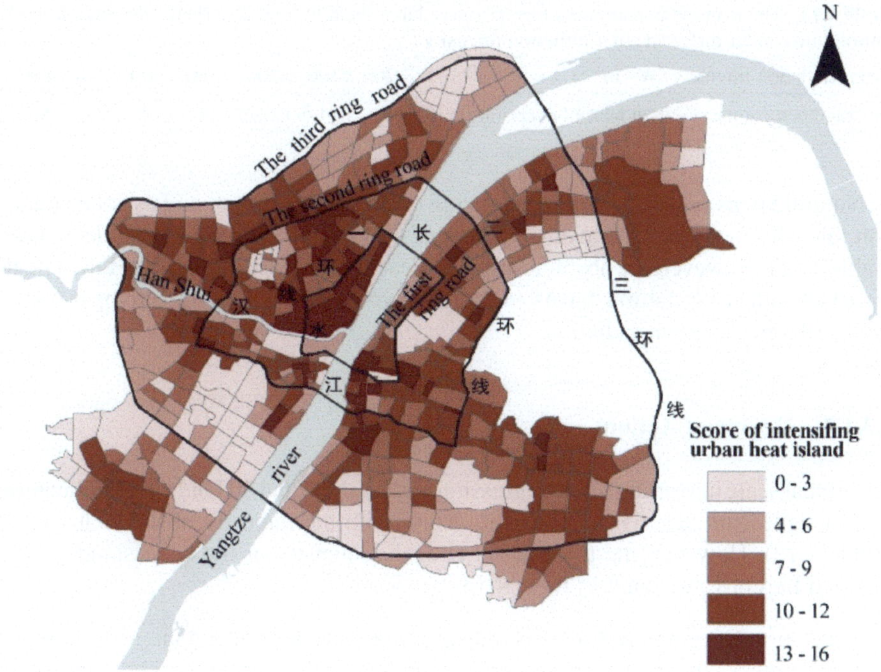

Fig. 4.10 The mechanism score of each regulation planning unit to increase heat island intensity

4.3 Mechanism of Urban Form Reducing Heat Island Intensity

4.3.1 Ecological Surface Heat Absorption Cooling: Ecological Surface Water Evaporation, Increasing Convective Heat Dissipation

4.3.1.1 Mode of Action

Various urban form indicators reflecting the ecological level, such as NDVI, green space rate, and vegetation coverage ratio, are significant factors affecting heat island intensity. Wet soil, vegetation, and water bodies lower their own temperature through evaporation, thereby reducing the surrounding heat accumulation. Trees not only block shortwave solar radiation but also interfere with longwave radiation from the ground, building surfaces, and the sky, thus reducing the sensible heat flux of the underlying surface [2]. Greening surfaces, including trees, grasslands, roof greening, and vertical greening, help avoid high temperatures through water evaporation. At 25 °C, the latent heat of water evaporation is 44 kJ/mol, the highest of all known solvents.

Table 4.11 The average percentage of evaporative latent heat in total absorbed solar radiation of natural lawn under different solar radiation intensity

Solar radiation intensity ($W \cdot m^{-2}$)	200	300	400	500	600	700	800
Average percentage of latent heat of evaporation /%	19	14.5	12.2	11	10	9	8.5

Natural lawns produce latent heat during evaporative cooling, and as solar radiation intensity increases, the average percentage of evaporative latent heat decreases (Table 4.11). However, the absolute value of heat absorbed by evaporative latent heat increases, allowing natural lawns to maintain relatively low surface temperatures even under higher solar radiation.

4.3.1.2 Degree of Action

The global scale univariate linear regression results show that all land cover indicators have a significant negative linear relationship with the intensity of the heat island (Table 4.12). However, the association with the intensity of the heat island varies across different value ranges (Table 4.13):

- Vegetation Coverage Ratio: This indicator mitigates heat island intensity in most value ranges. In regulatory planning units where the vegetation coverage ratio is less than 5%, the mitigating effect is the strongest, with heat island intensity increasing by 1.32 °C for every 10% increase in the proportion of industrial land.
- Green Space Rate: The impact of the green space rate on heat island intensity is most notable in areas where the proportion is less than 5% and more than 35%. Both ranges exhibit a mitigating effect, though the effect is significantly less than that of vegetation coverage.
- Soil Cover Ratio: In areas with soil cover ratios less than 5%, this indicator significantly mitigates heat island intensity, decreasing it by 1.51 °C for every 10% increase in the soil cover ratio.
- NDVI: The effect of NDVI on heat island intensity is primarily observed in areas where NDVI is greater than 0.4. Here, an increase of 0.1 in NDVI reduces heat island intensity by at least 0.70 °C.

Table 4.12 Linear regression coefficient of urban form indicators and urban heat island intensity of each core based on 589 management units

Core urban form indicators	Linear regression coeffficient
Ratio of green space (%)	-3.386^{***}
The proportion of vegetation cover (%)	-4.014^{***}
The proportion of soil cover (%)	-1.762^{***}
NDVI	-6.845^{***}

Note ***, **, * means passing 1, 5 and 10% significance level test respectively

Table 4.13 One-dimensional linear regression results of urban form indicators and urban heat island intensity in different value intervals

Indicator name and unit of measurement	The value range of the indicator	Linear regression coeffficient
The proportion of vegetation cover (%)	Less than 5	−13.219**
	5–10	−7.009
	10–20	−8.634**
	More than 20	−3.928***
Ratio of green space (%)	5%	−8.350*
	5–10%	9.096
	10–20%	−6.591
	20–35%	−2.902
	More than 35%	−2.867*
The proportion of soil cover (%)	5	−15.098**
	5–10	9.347
	10–20	−7.738
	20–30	−5.986
	More than 30	−1.981
NDVI	0.3	17.391**
	0.3–0.35	−9.260
	0.35–0.4	−2.086
	0.4–0.5	−8.155***
	More than 0.5	−6.968***

Note In parentheses are non-standard regression coefficients; *Note* ***, **, * means passing 1, 5 and 10% significance level test respectively

The statistical results indicate that surface coverage indicators generally exhibit a significant negative linear correlation with heat island intensity across their value ranges. When the ecological level in a management unit is extremely low, even small-scale ecological improvements can significantly reduce heat island intensity. This is likely because buildings and impervious surfaces dominate such areas, and a small increase in green surfaces enhances evaporative cooling effects while replacing heat-absorbing surfaces. Conversely, in regulatory planning units with a high ecological level, the cooling effect of ecological surfaces predominantly reduces heat island intensity. Their proportion can consistently influence heat island intensity.

4.3.2 Building Shadow Cooling: Reduced Openness Leads to Increased Shadow Area, Reducing Radiant Heat Absorption

When GSVF is high, the building combination exhibits good openness, resulting in a smaller building shadow area and a larger surface area exposed to solar radiation. Conversely, a low GSVF indicates that high-rise buildings may create extensive shadows, thus reducing the effective solar radiation interface (Fig. 4.11). In environments where the ground is predominantly covered by impervious surfaces, the extent of shadowing can significantly influence heat island intensity. Previous studies confirm that GSVF can mitigate heat island intensity in high-density urban areas [3, 4], while other research indicates that low-density residential areas may have a stronger heat island effect compared to high-density areas [5]. Therefore, GSVF affects heat island intensity in a dual manner: it can either mitigate heat island effects by promoting ventilation or intensify them by increasing radiant heat absorption from exposed surfaces.

From this perspective, the impact of GSVF on heat island intensity is dual. On one hand, lower GSVF hinders convective cooling, potentially increasing heat island intensity. On the other hand, higher GSVF exposes a greater area of hardened surfaces to direct sunlight, which can enhance radiant heat absorption and intensify the heat island effect. Therefore, the impact of GSVF on heat island intensity depends on external conditions: when vegetation predominates the land surface cover, the effect of building shadows on radiant heat absorption is minimal, and heat island intensity is primarily influenced by ventilation performance. In this case, areas with higher GSVF exhibit lower heat island intensity. Conversely, when the land surface is predominantly hardened, higher GSVF increases the exposure of surfaces to solar radiation, thus raising the heat island intensity.

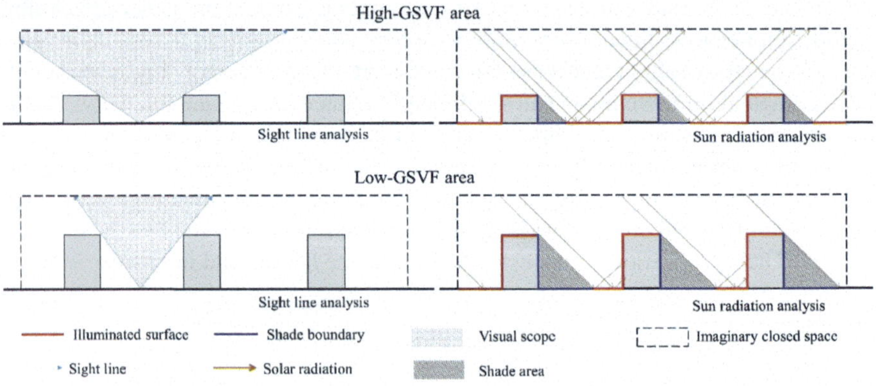

Fig. 4.11 Line-of-sight analysis (left) and solar radiation analysis (right) of building combinations with different GSVF values

In summary, GSVF plays a dual role in heat island intensity. Higher GSVF values can mitigate heat island intensity by enhancing ventilation through wind channels, but they can also increase heat island intensity by exposing more hardened surfaces to solar radiation due to a smaller building shadow area.

4.3.3 The Effect of Different Urban Forms in Reducing Heat Island Intensity

The aforementioned two types of impacts on heat island intensity both reflect the ability of ventilation and heat dissipation. Consistent with the previous text, each management unit is scored based on the mechanisms of these urban form indicators on heat island intensity (Table 4.14). Increasing the green space ratio, vegetation coverage ratio, soil coverage ratio, and NDVI enhances the overall heat dissipation potential of surface moisture, so higher values correspond to lower scores. Conversely, GSVF mitigates heat island intensity, with lower openness in building combinations creating larger shadows and reducing radiant heat absorption. Therefore, higher GSVF values also correspond to lower scores.

After summing the scores of the indicators that enhance the heat load, a spatial effect map of urban form indicators reducing heat island intensity is obtained. As shown in Fig. 4.12, the spatial distribution of mitigation mechanism scores lacks significant layer differentiation, with high- and low-score areas interspersed. The lowest scores (−13 to −16 points) are concentrated in certain areas at the edge of the study area. Scores in Hanyang and Wuchang are generally lower than those in Hankou, with continuous high-score areas along the Yangtze River in Hankou indicating a poor mitigating effect of the urban form on heat island intensity in this region.

4.4 Comprehensive Mechanism of Urban Form on Heat Island Intensity

4.4.1 Identification of External Environmental Factors

4.4.1.1 Spatial Distribution of Regression Residuals

To examine the accuracy of the regression of core urban form indicators on heat island intensity, we construct four OLS models (Table 4.15) using core urban form indicators from functional development, land surface coverage, construction intensity, and building combination as explanatory variables, with the heat island intensity of the management unit as the dependent variable. We then calculate the residuals for each management unit, which represent the difference between the observed and

Table 4.14 The scores of urban form indicators that have a moderating effect on heat island intensity

Urban form indicators	Value interval	Valuation
Ratio of green space (%)	More than 30%	−3
	10–30%	−2
	0–10%	−1
	0%	0
The proportion of vegetation cover (%)	More than 50%	−4
	25–50%	−3
	10–25%	−2
	0–10%	−1
	0%	0
The proportion of soil cover (%)	More than 25%	−3
	10–25%	−2
	0–10%	−1
	0%	0
NDVI	More than 0.6	−4
	0.5–0.6	−3
	0.4–0.5	−2
	0.3–0.4	−1
	Less than 0.3	0
GSVF (cooling mechanism by increasing building shadows)	Less than 0.80	−4
	0.80–0.85	−3
	0.85–0.90	−2
	0.95–0.95	−1
	More than 0.95	0

Note There is a certain similarity between the index of green land rate and the proportion of vegetation cover, and there are few literatures on the influence of supporting soil cover on heat island intensity, so the value is only divided into 3 grades

simulated values of heat island intensity. This analysis aims to identify potential control variables for inclusion in the model by examining the spatial distribution of the residuals.

The amount of the residuals reflects the accuracy of the simulation, while the sign of the residuals indicates whether the simulation results are higher or lower than the observed values. Based on the residuals, regulatory planning units are categorized into three groups: units with residual values greater than 0.5 °C are "areas where the simulation results are low," indicating that the simulation results are significantly lower than the observed values; units with residual values between −0.5 °C and 0.5 °C are "areas where the simulation results are ideal," indicating that the simulation results are close to the observed values. Units with residual values less than −0.5 °C are

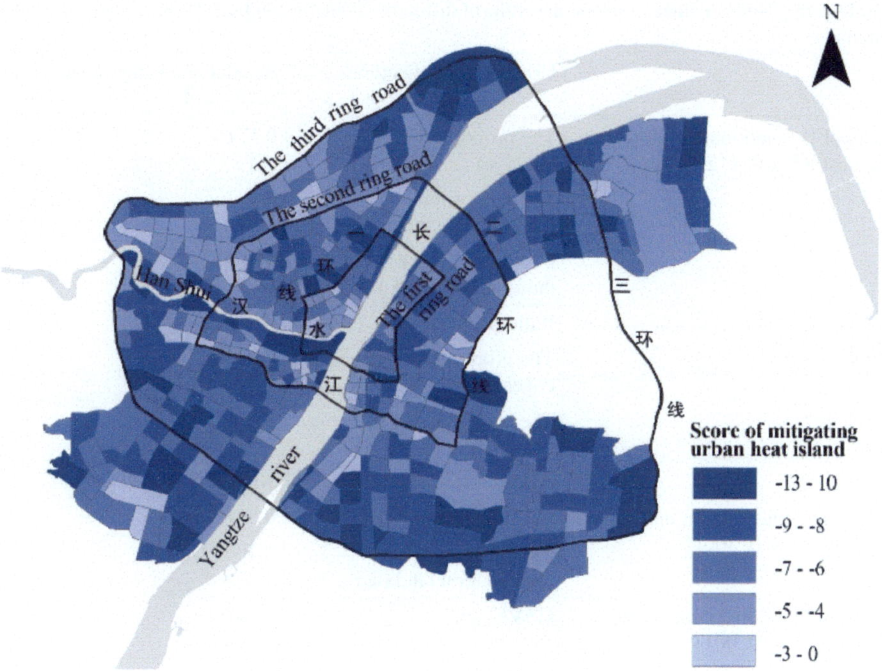

Fig. 4.12 Score of regulatory planning units regarding to reducing urban heat island

"areas where the simulation results are high," indicating that the simulation results are significantly higher than the observed values.

The spatial distribution of residuals reveals that different urban form indicators explain heat island intensity to varying extents across different regions (Figs. 4.13, 4.14, 4.15 and 4.16). Analysis of the residual distribution maps shows that: ① Functional development indicators are most effective at explaining heat island intensity, accounting for most regulatory planning units within the first ring and nearly one-third of the area between the second and third rings; ② Land surface coverage indicators best explain heat island intensity within the third ring of the Hanyang area; ③ Construction intensity indicators effectively explain heat island intensity along both sides of the Han River; ④ Building combination indicators explain heat island intensity in a more scattered fashion, covering parts of the Hanyang area within the third ring and some areas between the second and third rings of Hankou.

At the same time, the heat island intensity in some specific areas is often underestimated or overestimated by different urban form elements. For example, areas with concentrated industrial functions such as Zhuankou Industrial Zone, Wugang Industrial Zone, Gutian Industrial Zone, and Baisha Industrial Zone are often classified as "underestimated simulation results areas," where the simulation results fall short of observed values. Conversely, areas around Nanhu, Shahu, Moshui Lake, and South Taizi Lake are generally classified as "overestimated simulation results areas,"

Table 4.15 Multiple linear regression results of different urban form indicators on urban heat island intensity

	Urban form indicators	Standardized regression coefficient
Model 1: functional development (adjust $R^2 = 0.412$)	The proportion of industrial land (%)	0.574^{***}
	The proportion of commercial land (%)	0.187^{***}
	Population density (ten thousand people/km^2)	0.413^{***}
Model 2: land cover (adjust $R^2 = 0.337$)	Ratio of green space (%)	-0.162^{***}
	The proportion of vegetation cover (%)	-0.344^{***}
	The proportion of soil cover (%)	0.016
	NDVI	-0.337^{***}
Model 3: construction intensity (adjust $R^2 = 0.360$)	Building density	0.521^{***}
	The proportion of construction land (%)	0.106^{**}
Model 4: building combination (adjust $R^2 = 0.001$)	GSVF	−0.057

Note ***, **, * means passing 1, 5 and 10% significance level test respectively

where the simulation results exceed observed values. Additionally, spatial autocorrelation analysis reveals that the residuals from the simulations of the four urban form indicators exhibit significant clustering patterns (Table 4.16). Among these, the residuals from the construction intensity simulation show the highest degree of clustering, with a Moran's I index of 0.35. Further local spatial autocorrelation analysis (Anselin Local Moran's I) indicates that "high-high" clustering of residuals is typically found near major industrial areas, while "low-low" clustering occurs around significant water bodies.

4.4.1.2 Identification of Control Variables

Selection of Control Variables

Through the local spatial autocorrelation coefficient of residuals (Local Moran's I), the spatial distribution characteristics of residuals are reflected, and the potential urban form elements affecting the association between urban form indicators and heat island intensity are obtained, including ① concentrated industrial areas, ② main water body boundaries, ③ city trunk roads, ④ city ring roads, ⑤ city center (Fig. 4.17). Subsequently, the distances from the centroids of each management unit to these urban spatial elements are taken as explanatory variables, and the residuals of the

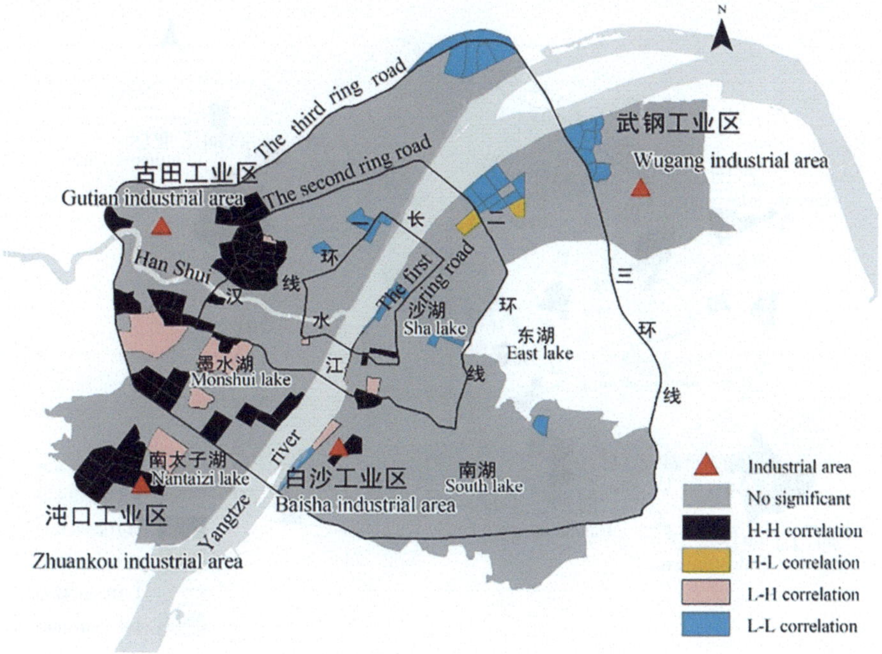

Fig. 4.13 Spatial autocorrelation of urban function model's residuals

aforementioned four linear regression results are taken as dependent variables to construct four linear regression models (Table 4.17), analyzing the impact of these urban form elements on the residuals.

The regression analysis results show that distances from concentrated industrial areas, main water body boundaries, and city trunk roads all have a significant and stable impact on the residuals in the four OLS models (Table 4.18). The regression coefficients for distances from concentrated industrial areas and trunk roads are negative, indicating that closer proximity to these features correlates with higher residuals, meaning that observed heat island intensities are much greater than those simulated based on urban form indicators, thus leading to an underestimation of actual heat island intensity in the simulation results. Conversely, the regression coefficients for distances from main water body boundaries are positive, indicating that closer proximity to these boundaries correlates with lower residuals, meaning that actual heat island intensities are much lower than the simulated values based on urban form indicators.

Contribution of Control Variables to the Model

The above analysis shows that the impact of various urban form elements on heat island intensity is influenced to varying degrees by external environmental factors.

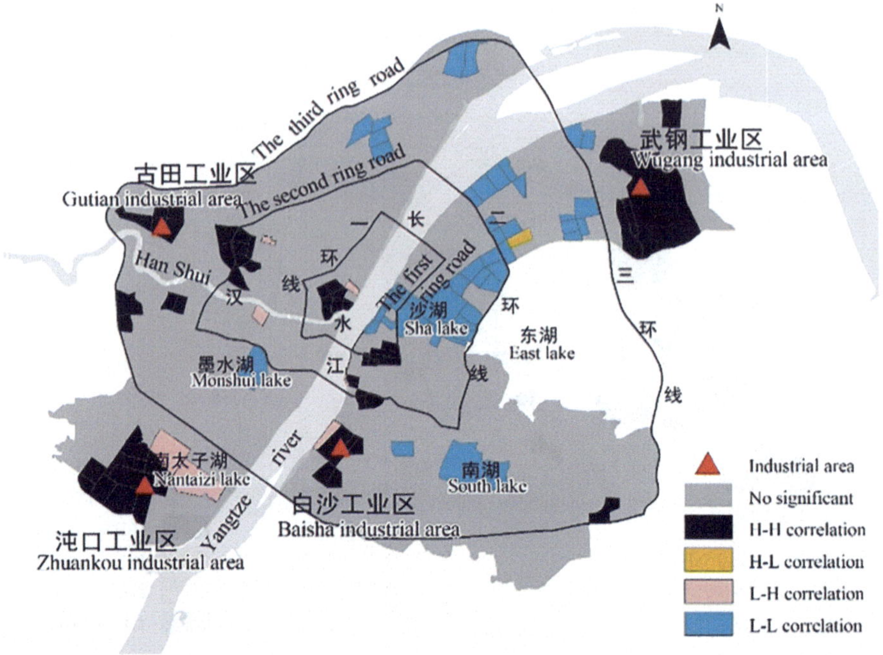

Fig. 4.14 Spatial autocorrelation of land cover model's residuals

In areas such as concentrated industrial zones, city trunk roads, and main water bodies, heat island intensity is significantly affected by industrial production, traffic, and water bodies, which weakens the impact of urban form elements on heat island intensity and complicates their explanation. After including the control variables of distance to the main industrial area, distance to the main water body boundary, and distance to the city's main roads in the multiple regression model, the explanatory power of the model slightly improved, with the adjusted R2 increasing from 0.678 to 0.718. Therefore, this book considers "distance to the main industrial area," "distance to the main water body boundary," and "distance to the city's main roads" as key external environmental factors affecting heat island intensity in relation to core urban form elements.

Spatial Effects of External Environmental Factors on Heat Island Intensity

Based on the above analysis of the impact of various external environmental factors on heat island intensity, each management unit is assigned values according to its distance from concentrated industrial areas, main water bodies, and main roads (Table 4.19). Concentrated industrial areas and main roads intensify the heat island effect in their surroundings, so the closer the distance, the stronger the intensification effect, and the higher the assigned value. Conversely, main water bodies alleviate the

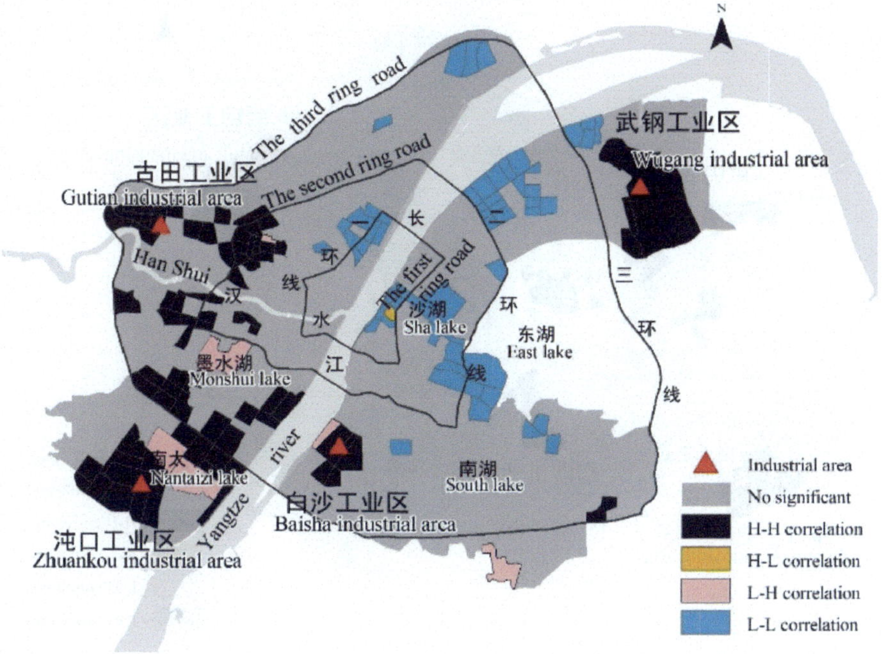

Fig. 4.15 Spatial autocorrelation of building intensity model's residuals

heat island effect in their surroundings, so they are assigned negative values, with closer distances resulting in smaller assigned values. After assigning values to the external environmental factors for each management unit, it is evident that these factors only alleviate heat island intensity in a few units located around the main lakes, while most other units experience intensified heat island effects due to external environmental factors (Fig. 4.18). Notably, the areas most affected by these factors are situated around the second ring road in the northwest, around the Wugang and Zhuankou industrial areas, and along the southeast side of the second ring road. This is primarily because, although the study area contains several large water bodies, including lakes and the Yangtze River, which help reduce heat island intensity, these specific areas are far from water bodies and therefore less influenced by their cooling effects. Instead, these areas are either close to concentrated industrial areas or have high road density, which significantly intensifies heat island effects.

4.4.2 Comprehensive Impact and Spatial Effects

Based on the analysis, factors affecting heat island intensity can be categorized into three main types: intensifying factors that contribute to warming, mitigating factors

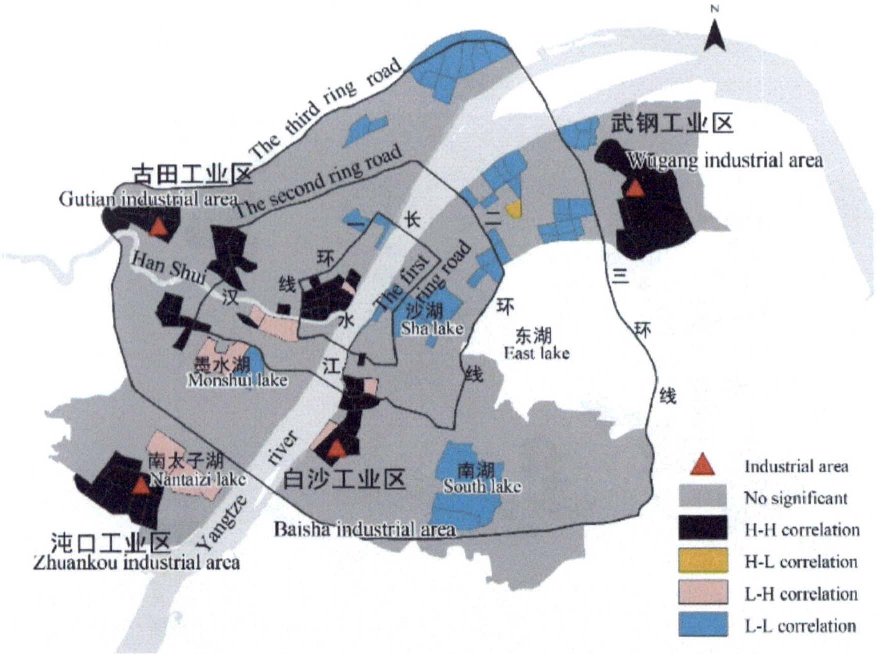

Fig. 4.16 Spatial autocorrelation of building combination model's residuals

Table 4.16 Spatial autocorrelation coefficient of residuals of regression results of different urban form elements

	The Moran I index	Z score	P value
Simulated residual differences of functional development	0.27	13.86	0.00
Simulated residuals of the ground surface cover	0.24	12.34	0.00
Simulated residuals of the construction intensity	0.35	18.25	0.00
Simulated residuals of the building combinations	0.29	15.33	0.00

Note The default domain search threshold is 1668.45 m

that contribute to cooling, and external environmental factors. The interaction of these factors creates high and low variations in heat island intensity. This book scores each management unit based on these factors to reflect their influence on heat island intensity. Industrial land, commercial land, permanent population density, building density, and construction land uniformly intensify heat island effects, while green space rate, vegetation coverage, soil coverage, and NDVI uniformly mitigate heat island effects. The effect of GSVF is complex and varies with its value; a GSVF less than 0.95 hinders convective heat dissipation, but a smaller GSVF also increases building shadows that can reduce solar radiation and heat island intensity. Water bodies mitigate heat island effects, while main roads and industrial areas intensify

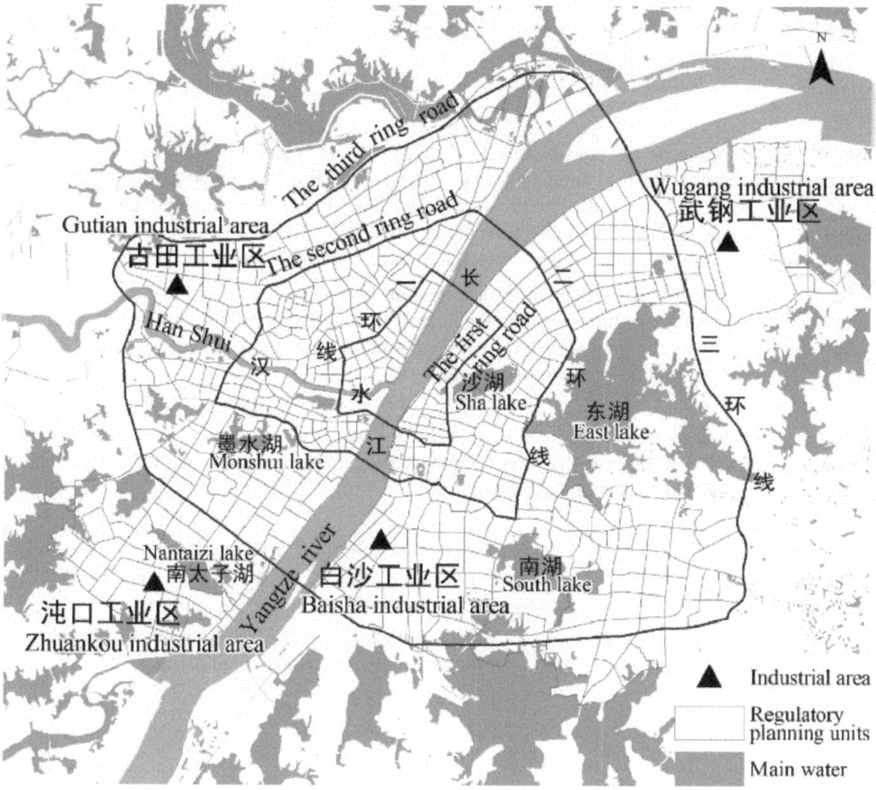

Fig. 4.17 Major industrial area, road and water

them. By scoring each management unit, the spatial distribution of the overall effect of urban form on heat island intensity can be illustrated. The combined effects of these factors determine the magnitude of the impact of urban form indicators on heat island intensity (Fig. 4.19).

After aggregating the scores for the strengthening effect of urban form indicators on heat island intensity, the mitigating effects, and the location factors, the comprehensive impact of urban form indicators on heat island intensity is determined. The comprehensive score for each management unit reflects the magnitude of this impact. Among the 589 regulatory planning units, the comprehensive scores range from −14 to 16. A higher score indicates a stronger strengthening effect and a weaker mitigating effect on heat island intensity, signifying greater heat load and reduced ventilation capacity. Conversely, a lower score indicates a stronger mitigating effect and weaker strengthening effect, reflecting a lower heat load and enhanced ventilation capacity (Fig. 4.20).

To test the explanatory power of the comprehensive score on actual heat island intensity, the average absolute difference and correlation coefficient were calculated, resulting in values of 0.337 and 0.637, respectively. The low average absolute

Table 4.17 Main spatial variable index and its measuring method

Spatial variable name	Measuring method
Distance from the main industrial area	According to the construction land status map, choose Zhuankou industrial zone, Wugang industrial zone, Gutian industrial zone, Baisha industrial zone geometric center, and then calculate the distance between the center of mass of each management unit and the above main industrial geometric center, select the minimum distance, and assign values to each management unit
Distance from the main water body boundary	The water body with an area of more than 5 hectares is selected to extract the water body boundary line, and then the nearest distance between the centroid of each management unit and the water body boundary line is calculated, and each management unit is assigned a value
Distance from the urban trunk road	The center line of the main road in the research area is extracted, the nearest distance of the center of mass of each management unit is measured, and the value is assigned to each management unit
Distance from the city loop	Extract the first ring line, second ring line and third ring line of Wuhan city, measure the nearest distance between the centroid of each management unit and them, and assign a value to each management unit
Distance from the center of the main city area	Extract the overall geometric center of the study scope, calculate the linear distance of each management unit, and assign value to each management unit

error and the high correlation coefficient indicate a strong correlation between the comprehensive score and the observed heat island intensity. This demonstrates that the comprehensive score effectively predicts heat island intensity, supporting the validity of the proposed comprehensive impact mechanism.

4.4.3 Comparison of the Impact of Various Urban Form Indicators

To compare the impact of different core urban form indicators on heat island intensity, a multiple linear regression model was constructed using 10 core urban form indicators as explanatory variables, environmental impact factors as control variables, and heat island intensity as the dependent variable. This approach generated multiple core urban form indicator models for heat island intensity (Table 4.20). Compared to univariate linear regression, multiple linear regression offers a more comprehensive view of the relationships between heat island intensity and various urban form indicators. It allows for the examination of how changes in one explanatory variable affect the dependent variable while holding other variables constant,

Table 4.18 The regression coefficient of different spatial variables to the residual of simulation results based on different urban form elements

	Model 1: residue regression based on functional development elements	Model 2: residue regression based on surface cover elements	Model 3: residual regression based on construction intensity elements	Model 4: residue regression based on building combination elements
Distance from the main industrial area	−0.154***	−0.134***	−0.147***	−0.118***
Distance from the main water bodies	0.140***	0.122***	0.134***	0.108***
Distance from the trunk road	−0.102***	−0.088***	−0.097***	−0.078***
Distance from the city loop	−0.015	−0.013	−0.014	−0.011
Distance from the center of the main city area	−0.007	−0.006	−0.007	−0.005

Note Due to the large number of spatial form indicators, the regression results of each urban form index and heat island intensity in each model were not included in the table; ***, **, * means passing 1, 5, and 10% significance level tests respectively

Table 4.19 Effect score of external environmental factors on urban heat island intensity

Urban form indicators	Value interval	Valuation
Distance from the concentrated industrial area (km)	Less than 3	3
	3–6	2
	6–9	1
	More than 9	0
Distance from the main water body (km)	Less than 0.8	−3
	0.8–1.5	−2
	1.5–2.5	−1
	Greater than 2.5	0
Distance from the main road (km)	Less than 0.5	3
	0.5–1.0	2
	1.0–2.5	1
	More than 2.5	0

Fig. 4.18 Score of environmental impact on urban heat island

thereby capturing the collective impact of multiple urban form indicators on heat island intensity.

According to the linear regression results, the linear regression equation is: $Y = 1.743X_1 + 0.990X_2 + 0.157X_3 - 0.175X_4 + 0.428X_5 + 4.006X_6 - 3.616X_7 + 7.092X_8 + 1.505X_9 + 9.499X_{10} - 0.056X_M - 0.135X_R + 0.183X_W - 7.669$.

Where, Y: Heat island intensity; X_1: Proportion of industrial land; X_2: Proportion of commercial land; X_3: Permanent population density; X_4: Green space rate; X_5: Proportion of vegetation cover; X_6: Proportion of soil cover; X_7: NDVI; X_8: Building density; X_9: Proportion of construction land; X_{10}: GSVF; X_M: Distance to concentrated industrial area; X_R: Distance to main roads; X_W: Distance to the boundaries of various large water bodies.

The results of the multiple linear regression analysis revealed that building density and GSVF have the largest standard regression coefficients, indicating that they exert the greatest influence on heat island intensity among all urban form indicators. The proportion of covered land area, NDVI, proportion of construction land, and density of permanent residents follow, with the proportion of industrial land and commercial land having a comparatively smaller impact. Vegetation cover and green space rate do not show a significant effect on heat island intensity in the model. Overall, building related indicators have the highest impact on heat island intensity, followed by surface cover indicators, while land use indicators have the least impact. The influence of

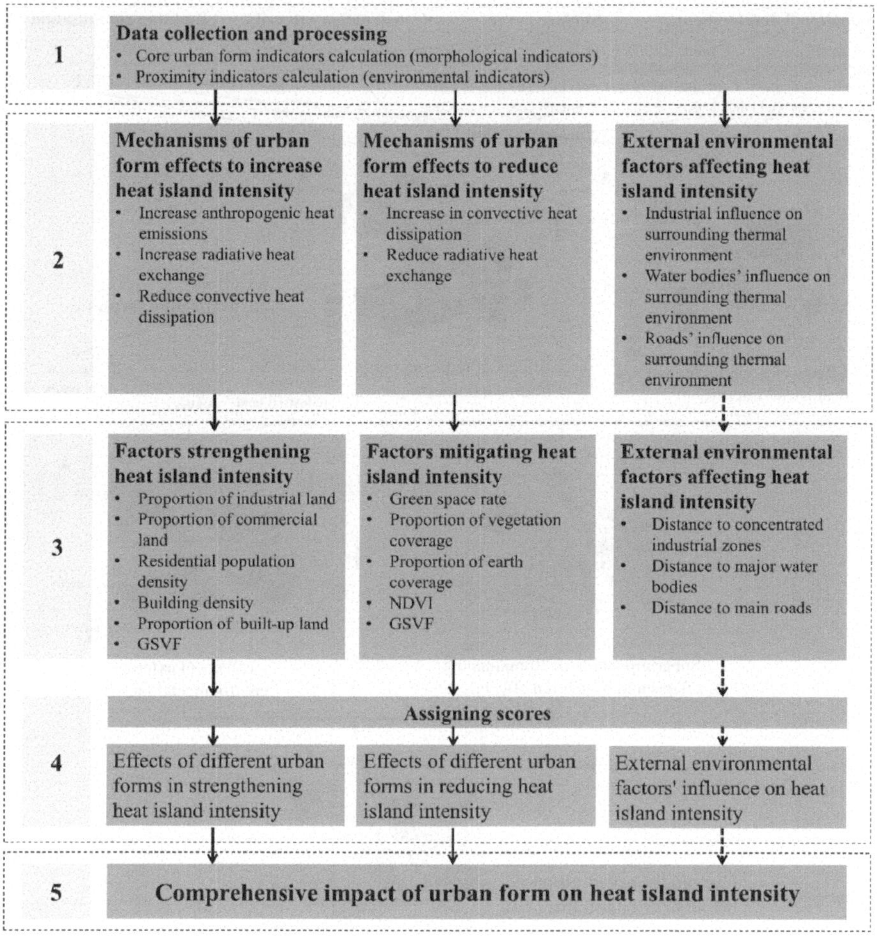

Fig. 4.19 Analysis framework of urban form impact on urban heat island

vegetation cover and green space rate on heat island intensity is largely replaced by NDVI, due to high collinearity between them. Although GSVF does not exhibit a significant linear correlation with heat island intensity in univariate regression, it becomes significant in the multiple linear regression, indicating that increased GSVF can elevate heat island intensity when other urban form indicators are held constant.

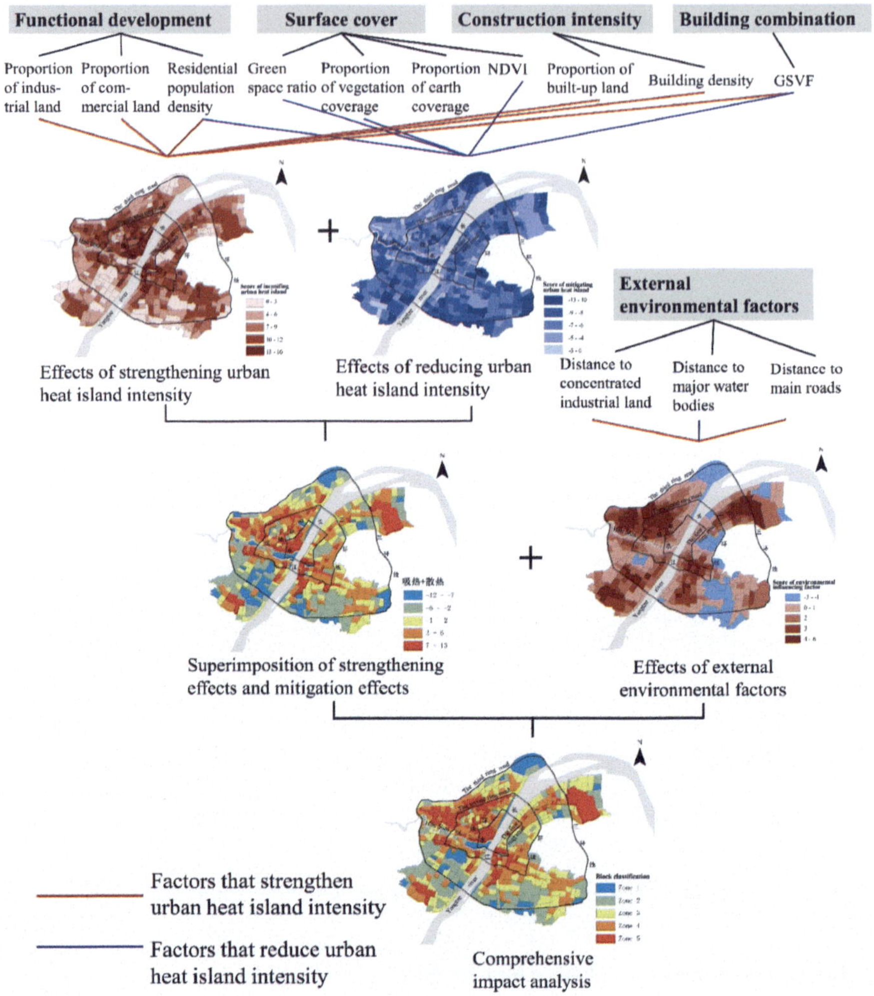

Fig. 4.20 Flow chat of urban form impact on urban heat island

4.5 Summary

After selecting 10 core urban form indicators that affect heat island intensity in Chapter 4, this chapter focuses on the impact mechanisms of these indicators. First, a theoretical framework for these mechanisms was constructed, using the energy balance within the canopy space as the basis for understanding heat transfer. Three basic heat conduction methods were proposed: artificial heat dissipation, radiation heat gain, and convection heat dissipation. The urban form indicators were then linked to these methods, revealing that their impact on heat island intensity includes

Table 4.20 Regression results of OLS model between core urban form indicators affecting heat island intensity and heat island intensity

	Name of index	Variable code	Standardized regression coefficient
Explanatory variable	The proportion of industrial land (%)	X_1	0.196 (1.743)***
	The proportion of commercial land (%)	X_2	0.056 (0.990)**
	Population density (ten thousand people/km^2)	X_3	0.219 (0.157)***
	Ratio of green space (%)	X_4	−0.019 (−0.175)
	The proportion of vegetation cover (%)	X_5	0.046 (0.428)
	The proportion of soil cover (%)	X_6	0.321 (4.006)***
	NDVI	X_7	−0.233 (−3.616)***
	Building density	X_8	0.591 (7.092)***
	The proportion of construction land (%)	X_9	0.264 (1.505)***
	GSVF	X_{10}	0.489 (9.499)***
Control variable (as auxiliary correction)	Distance from the concentrated industrial area (km)	X_m	−0.139 (−0.056)***
	Distance from the main road (km)	X_r	−0.088 (−0.135)***
	Border distance from various large area water bodies (km)	X_w	0.131 (0.183)***

Note In parentheses are non-standard regression coefficients; ***, **, * means passing 1, 5, and 10% significance level tests respectively

strengthening mechanisms, slowing mechanisms, external environment correction mechanisms, and a comprehensive mechanism that combines these effects.

The strengthening mechanisms of urban form on heat island intensity include ‘functional operation heating’, ‘artificial surface heat exchange’, and ‘building combination heat accumulation’. Through quantitative correlation analysis, it was found that the proportion of industrial land, commercial land, and permanent population density promote the heating effect of functional operations while building density and the proportion of construction land affect the efficiency of artificial surface radiation heat gain. GSVF affects building combinations by slowing airflow and creating aggregated heat. The combined effect of these mechanisms constitutes the strengthening impact of urban form on heat island intensity, which was quantified through a scoring system for 589 regulatory planning units.

The book also addresses mechanisms that slow down heat island intensity, focusing on 'ecological surface heat absorption' and 'building combination heat reduction'. Quantitative correlation analysis showed that green space ratio, vegetation coverage, soil coverage ratio, and NDVI influence surface water evaporation, while GSVF determines the building shadow area. These factors together determine the slowing effect of urban form on heat island intensity, which was also quantified through a scoring system for 589 regulatory planning units.

The combined effects of the strengthening and slowing mechanisms, along with external environmental factors, were used to form a comprehensive mechanism for urban form indicators affecting heat island intensity. The corrective effects of external environmental factors account for the impact of concentrated industrial areas, major water bodies, and main roads on surrounding heat island intensity. The comprehensive impact of each urban form indicator on heat island intensity was assessed, and it was found that the comprehensive mechanism score aligns well with observed heat island intensity, indicating the validity of the mechanism analysis. Finally, a multiple linear regression model was used to compare the impacts of different indicators on heat island intensity, revealing that, except for green space and vegetation coverage, all other indicators significantly relate to heat island intensity, with building density having the greatest impact.

References

1. Ke TR (1987) Boundary layer climate. Methuan & Co. Ltd., London, p 274
2. Chen H, Li B, Zhang W (2015) Correlation between urban microclimate regulation and block morphological elements. Urban Architec 31:41–43
3. Middel A, Häb K, Brazel AJ et al (2014) Impact of urban form and design on mid-afternoon microclimate in Phoenix local climate zones. Landsc Urban Plan 122:16–28
4. Zhang H, Zhu SY, Gao Z, Zhang GX (2016) Relationship between urban morphological parameters and urban heat island intensity under clear weather conditions. J Appl Meteorol 27(02):249–256
5. Stone B Jr, Rodgers MO (2001) Urban form and thermal efficiency: how the design of cities influences the urban heat island effect. J Am Plan Assoc 67(2):186–198

Chapter 5
Urban Form Regulation Techniques for Reducing Heat Island Intensity

Based on the conclusions of the previous research, this chapter proposes three main urban form regulation techniques.

First, Urban form Indicator Warning and Optimization Techniques: This technique involves developing a warning indicator system to identify areas needing improvement. Specific urban form optimization strategies are then proposed based on this system to enhance the effectiveness of heat island mitigation efforts.

Second, Differentiated Guidance Techniques for Different Block Types: This approach provides tailored guidance strategies for various block types, taking into account the different correlations between urban form indicators and heat island intensity across these blocks. The strategies are designed to address the unique needs of each block type based on its specific characteristics and heat island impact.

Third, Urban form Optimization Techniques for Special Function Areas: This technique focuses on proposing urban form optimization measures for special function areas. These measures are aimed at addressing the specific spatial elements and functions of these areas to better manage their heat island effects.

Finally, a case study of Hualou Street is presented. Based on the specific conditions of Hualou Street, targeted land use regulatory measures are proposed to demonstrate the practical application of the proposed techniques.

Y. Lu, *Urban Form and Urban Heat Island*, Urban Sustainability,
https://doi.org/10.1007/978-981-96-3541-2_5

5.1 Urban Form Indicator Warning and Optimization Techniques

5.1.1 Threshold Analysis of Each Indicator Affecting the Formation of High-Temperature Areas

The proportion of regulatory planning units within the extreme heat island area and the strong heat island area across different value ranges for each core urban form indicator was statistically analyzed. The analysis revealed that when certain urban form indicators exceed specific values, the probability of a management unit falling into the extreme or strong heat island categories increases significantly. This specific value is referred to as the "threshold" for the core urban form indicator's impact on the formation of extreme and strong heat island areas.

The statistical results indicate that several core urban form indicators, including the proportion of industrial land, commercial land, permanent population density, vegetation coverage, soil coverage, building density, and construction land, exhibit a "threshold" effect on the formation of extreme and strong heat island areas. Out of 589 regulatory planning units, 48.39% are classified as extreme or strong heat island areas. Therefore, a threshold is defined as the maximum value within a "continuous numerical range" where the probability of forming an extreme or strong heat island area exceeds the overall average of 48.39%. If no such continuous value range exists, it is concluded that there is no discernible "threshold." Based on this method, the formation ratios of strong and extreme heat island areas across different ranges of each core spatial indicator are as follows: 13.

5.1.1.1 Proportion of Industrial Land

As the proportion of industrial land increases, the probability of forming extreme and strong heat island areas also increases (Fig. 5.1). When the proportion of industrial land is below 8%, the probability of forming strong and extreme heat island areas does not exceed the overall average. The probability fluctuates significantly between 8 and 24%, but once the proportion exceeds 26%, the likelihood of forming extreme and strong heat island areas rises sharply and stabilizes at 100% when the proportion exceeds 56%. Therefore, the critical value for the formation of extreme heat island areas and strong heat island areas is 24% industrial land.

5.1.1.2 Proportion of Commercial Land

The numerical range of the proportion of commercial land is also directly proportional to the probability of forming extreme heat island areas and strong heat island areas (Fig. 5.2). Statistical analysis reveals that when the proportion of commercial land

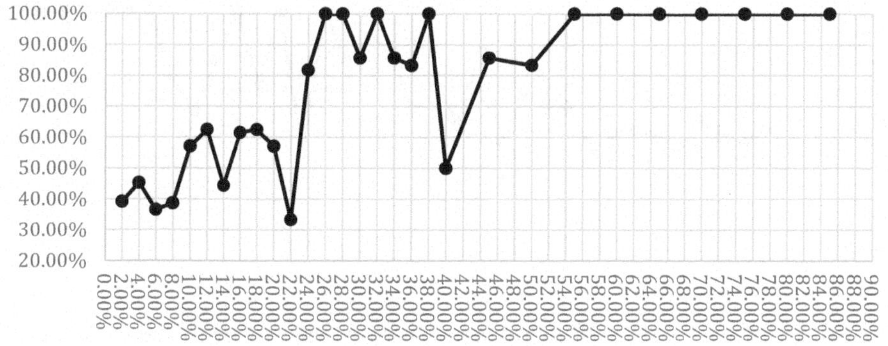

Fig. 5.1 Proportion of extreme urban heat island intensity zone and intensive urban heat island intensity zone in areas with different industrial area proportion

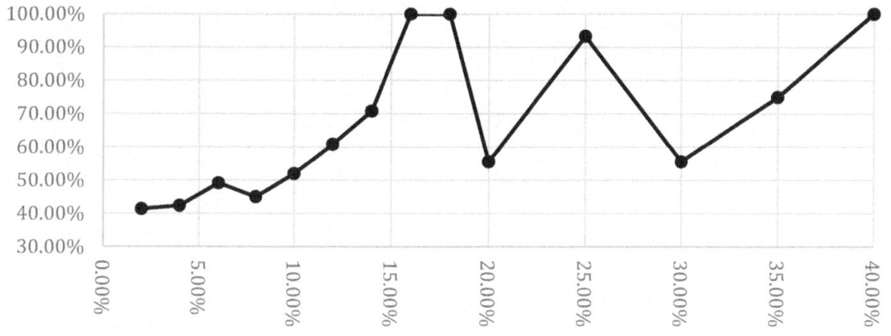

Fig. 5.2 Proportion of extreme urban heat island intensity zone and intensive urban heat island intensity zone in areas with different commercial area proportion

exceeds 10%, the probability of forming both extreme and strong heat island areas remains high across different ranges. Thus, a 10% proportion of commercial land is identified as the critical threshold for significantly increasing the risk of extreme and strong heat island effects.

5.1.1.3 Permanent Population Density

Permanent population density also plays a significant role in promoting the formation of extreme and strong heat island areas, with the overall trend indicating that higher population densities correspond to greater probabilities of extreme heat island conditions (Fig. 5.3). Statistical analysis reveals that when the permanent population density exceeds 30,000 people/km^2, the probability of forming extreme heat island areas and strong heat island areas is consistently higher than the average and shows an increasing trend. Furthermore, when the density surpasses 58,000 people/km^2, the

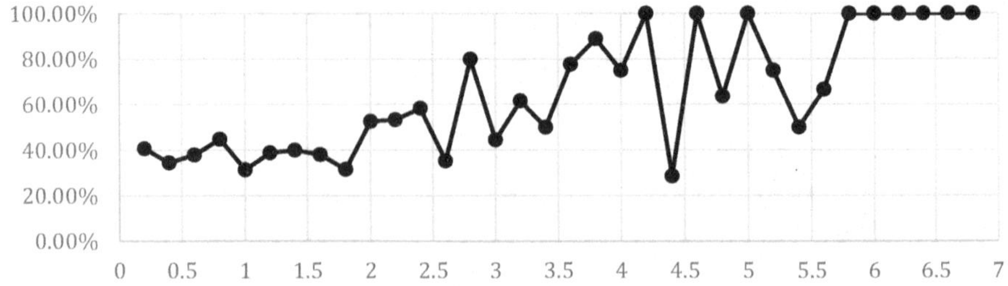

Fig. 5.3 Proportion of extreme urban heat island intensity zone and intensive urban heat island intensity zone in areas with different population density

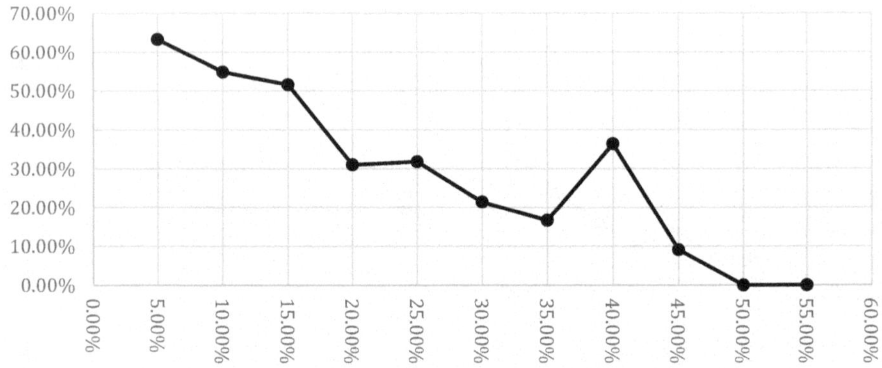

Fig. 5.4 Proportion of extreme urban heat island intensity zone and intensive urban heat island intensity zone in areas with different vegetation area proportion

probability reaches 100%, confirming that 30,000 people/km^2 is a critical threshold for the formation of extreme and strong heat island effects.

5.1.1.4 Proportion of Vegetation Coverage

The numerical range of the proportion of vegetation coverage is inversely proportional to the probability of forming extreme heat island areas and strong heat island areas (Fig. 5.4). In the numerical range less than 15%, the probability values are all greater than 50%, indicating that its critical value is 15%.

5.1.1.5 Proportion of Soil Coverage

The impact of soil coverage on heat island intensity does not show a clear threshold (Fig. 5.5).

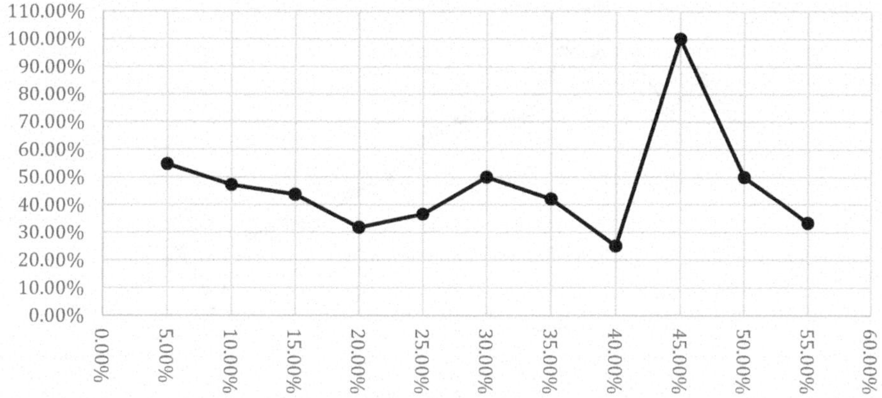

Fig. 5.5 Proportion of extreme urban heat island intensity zone and intensive urban heat island intensity zone in areas with different earthing area proportion

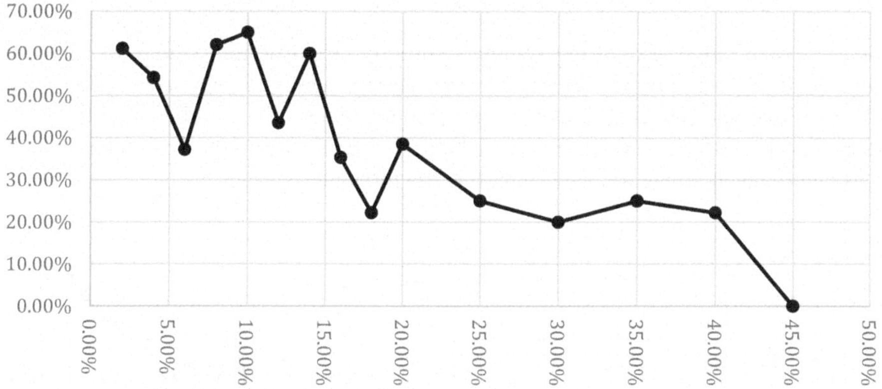

Fig. 5.6 Proportion of extreme urban heat island intensity zone and intensive urban heat island intensity zone in areas with different green area proportion

5.1.1.6 Greening Rate

The greening rate's impact on heat island intensity does not show a clear threshold (Fig. 5.6).

5.1.1.7 NDVI

The partial numerical range of NDVI (0.24–0.55) is inversely proportional to the probability of forming strong heat island areas and extremely strong heat island areas (Fig. 5.7). When NDVI is less than 0.42, the probability values in each interval stabilize at the overall average of 48.39%, so the threshold value is 0.42.

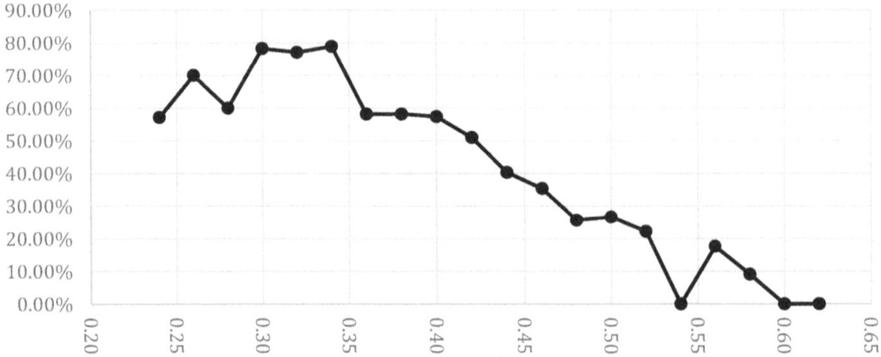

Fig. 5.7 Proportion of extreme urban heat island intensity zone and intensive urban heat island intensity zone in areas with different NDVI

5.1.1.8 Building Density

The numerical range of building density is directly proportional to the probability of forming strong heat island areas and extremely strong heat island areas (Fig. 5.8). When the average building density of the management unit is higher than 0.25 14, the probability of forming strong heat island areas and extremely strong heat island areas stabilizes at above 48.39%, which is the threshold value.

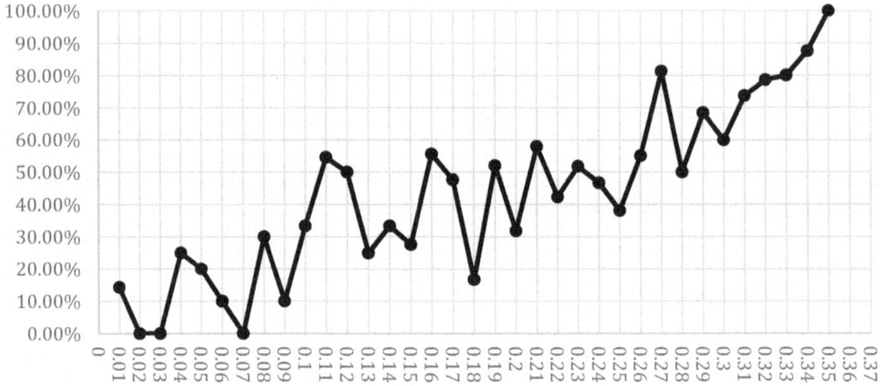

Fig. 5.8 Proportion of extreme urban heat island intensity zone and intensive urban heat island intensity zone in areas with different building density

5.1.1.9 Proportion of Construction Land

When the proportion of construction land exceeds 40%, its interval value is directly proportional to the probability of forming extremely strong heat island areas and strong heat island areas, with a threshold value of 70% (Fig. 5.9).

GSVF

The relationship between GSVF (Green-Sky View Factor) and heat island intensity is complex and nuanced. In previous Pearson correlation analyses, GSVF did not exhibit a significant correlation with heat island intensity across 589 regulatory planning units. This result supports the finding from the mechanism analysis that GSVF has both reinforcing and mitigating effects on heat island intensity. Statistical analysis indicates that when GSVF reaches approximately 0.85, the probability of forming extreme and strong heat island areas peaks, while the probabilities of such formations in areas with either very high or very low GSVF are comparatively unstable and generally lower (Fig. 5.10). This suggests that GSVF exerts its maximum reinforcing effect on heat island intensity at a specific threshold value: When GSVF is less than 0.85, the low sky visibility due to high-rise building shading limits the artificial surfaces exposed to solar radiation, thereby increasing heat island intensity. Conversely, when GSVF exceeds 0.85, the high sky visibility facilitates better ventilation and heat dissipation, reducing the likelihood of extreme heat island conditions. Therefore, the critical range for GSVF in influencing heat island intensity is approximately 0.84 to 0.87.

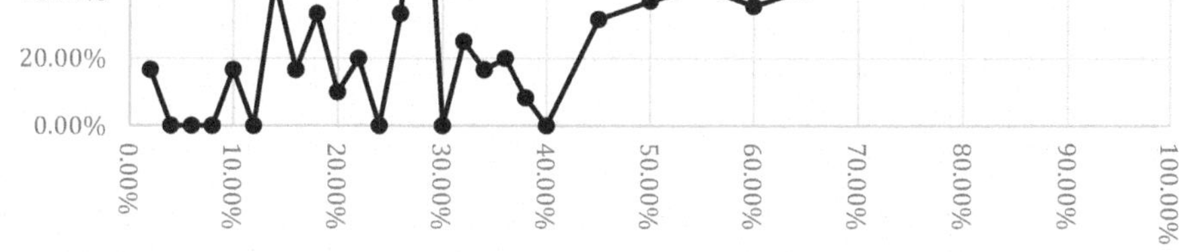

Fig. 5.9 Proportion of extreme urban heat island intensity zone and intensive urban heat island intensity zone in areas with different built-up land proportion

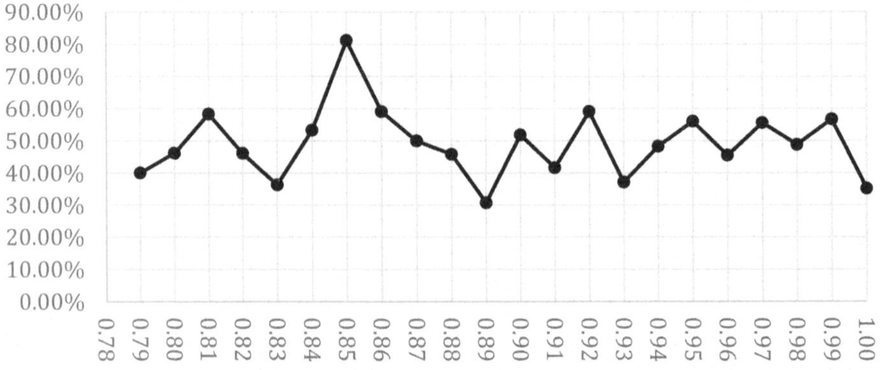

Fig. 5.10 Proportion of extreme urban heat island intensity zone and intensive urban heat island intensity zone in areas with different ground surface view factor

5.1.2 Indicator Warning and Regulation Measures

5.1.2.1 Industrial Land Regulation

Threshold analysis reveals that areas with more than 24% industrial land exhibit a consistently high probability of developing strong heat island conditions. Based on this threshold, three strategic approaches for the future functional development of these areas are proposed to mitigate the heat island effect and optimize land use:

- Reduce Industrial Land Proportion: Lower the proportion of industrial land to below 24% and replace it with green spaces or other non-industrial land uses. This strategy aims to decrease the likelihood of forming strong heat island areas by increasing the proportion of land that contributes to cooling rather than heating.
- Relocate Residential and Public Services: Move residential and public service functions away from high-intensity heat island areas. By relocating these functions to areas with lower heat island intensity, the exposure of daily activity areas to extreme heat conditions can be minimized.
- Optimize Spatial Relations: Maintain the current proportion of industrial land but improve spatial planning to better separate industrial activities from residential areas. This can be achieved through functional adjustments and the creation of green belts to act as buffers, thus reducing the direct impact of industrial land on residential areas (Table 5.1).

From a land economic perspective, the strategy involves gradually relocating existing industrial activities from the central urban area to suburban locations where they can be concentrated on a larger scale. This approach not only alleviates the heat island effect but also capitalizes on the higher value of the vacated land in the urban core, which can be repurposed for higher-rent uses. Consequently, the

Table 5.1 Industrial land control strategy

The current situation index range of the early warning object	Control method
Less than 24%	–
More than 24%	Reduce the proportion of industrial land to below the threshold level of 24%; Relocation of residential land out of industrial land accounts for a large number of management units; The construction of green separation between concentrated industrial land and residential land

restructured industrial zones in the suburbs will function as concentrated industrial areas, minimizing their disruptive effects on residential neighborhoods.

5.1.2.2 Commercial Land Regulation

To mitigate the adverse impacts of commercial land on the thermal environment of residential areas, two main strategies are proposed.

In areas where the proportion of commercial land exceeds 10%, it is essential to integrate ecological features such as water bodies and green spaces into the urban planning framework. This approach involves concentrating commercial facilities to minimize the direct interaction between commercial and residential zones. By creating physical and functional buffers between these areas, such as parks or green belts, the intensity of heat islands generated by commercial activities can be reduced. This strategy aims to moderate the effects of high-intensity heat islands on adjacent residential areas by focusing commercial functions in designated zones and minimizing their impact on nearby residential environments.

In areas where the proportion of commercial land is currently less than 10%, it is important to be vigilant about the potential for future commercial expansion. Strategies should include regulating the growth of commercial activities to keep their proportion under 10% and relocating commercial functions that are currently mixed with residential uses. This regulation aims to prevent the increase in artificial heat emissions from commercial functions that could negatively impact residential areas. By controlling commercial land use, the heat island effects associated with commercial activities can be managed more effectively (Table 5.2).

From a land economic perspective, concentrating on commercial functions can meet the spatial needs of diverse commercial formats. Commercial agglomeration creates an environment that reduces the costs of multi-target shopping trips for consumers, thus enhancing the attractiveness of commercial areas. The spatial concentration of commercial enterprises can generate economic benefits and foster positive externalities, such as increased foot traffic and consumer convenience [1].

Table 5.2 Commercial land control strategy

The current situation index range of the early warning object	Control method
<10%	The commercial functions mixed in the residential area will be relocated to reduce the influence of artificial heat removal of commercial functions on residential functions; The proportion of commercial land will be controlled within 10%
≥10%	Encourage the centralized layout of commercial facilities to avoid excessive mixing of residential and commercial

5.1.2.3 Permanent Population Density Regulation

Threshold research indicates that when permanent population density exceeds 30,000 people/km^2, heat island intensity stabilizes at a high level. Regulatory planning units with a permanent population density above 30,000 people/km^2 are predominantly located in the historic districts of Hankou and Wuchang, often characterized by aging residential neighborhoods. In extreme cases, the permanent population density in some units exceeds 100,000 people/km^2. To mitigate the impact of high population density on heat island intensity, it is essential to advance the transformation of these old neighborhoods into modern residential communities that adhere to current technical standards. This approach not only reduces population density but also lessens solar radiation through the mutual shading provided by high-rise buildings. For areas where the current permanent population density is below 30,000 people/km^2, it is crucial to manage future population growth by controlling residential construction and maintaining a population density under 30,000 people/km^2 (Table 5.3).

Table 5.3 Population density control strategy

The current situation index range of the early warning object	Control method
<30,000 people/km^2	The permanent population density will be controlled below 30,000 /km^2 to avoid rapid population density growth
≥300,000 people/km^2	To replace the existing "high density, low height" old residential areas with "low density" modern residential areas, so as to reduce population density

5.1.2.4 Land Surface Cover Indicator Regulation

Among the core urban form indicators affecting heat island intensity, the proportion of vegetation coverage, NDVI, green space rate, and soil coverage area are all surface coverage indicators. Among these, the proportion of vegetation coverage and NDVI have a larger relationship with heat island intensity and exhibit a threshold effect, while the green space rate and soil coverage proportion have a smaller impact. A comparison of the calculation principles of these indicators reveals that although vegetation coverage, NDVI, and green space rate all reflect ecological levels, their focuses are distinct. The green space rate measures the proportion of relatively pure green areas, such as parks and protective green spaces, excluding green space coverage in other land types. In contrast, vegetation coverage and NDVI encompass all types of land use, including green spaces in non-green construction areas, such as central green spaces in residential zones and attached green spaces.

The proportion of vegetation coverage is significantly associated with heat island intensity across various value ranges, with the regression coefficient being largest when vegetation coverage is less than 5%, reaching −13.219. Although vegetation coverage generally alleviates heat island intensity, its effect is particularly critical in areas with scarce vegetation. Threshold analysis reveals that when vegetation coverage is below 15%, the probability of forming high-intensity heat island areas is both higher and more stable. Such areas are often characterized by high building density, extensive hardened surfaces, and a lack of ecological resources, commonly found in large-scale industrial zones and old urban districts where it is difficult to establish extensive green spaces. In these contexts, every opportunity should be utilized to increase vegetation coverage through measures such as creating road green belts and adding green spaces to buildings, thereby enhancing three-dimensional greening and adopting a "needle-threading" approach to improve overall vegetation coverage. For areas where vegetation coverage is currently above 15%, it is important to prevent further loss of vegetation coverage in future urban development to maintain the proportion above this threshold (Table 5.4).

Table 5.4 Vegetation cover control strategy

The current situation index range of the early warning object	Control method
<15%	Use all available sites to increase vegetation coverage, such as road green belt, building attached green space, etc., to increase the proportion of vegetation coverage above the threshold level of 15%, such as increasing the ground green area, adding new park green space, protective green space, and improving the level of three-dimensional greening
≥15%	Beware of excessive loss of vegetation cover in future urban development, and maintain the proportion of vegetation cover at more than 15%

Table 5.5 NDVI control strategy

The current situation index range of the early warning object	Control method
<0.42	Increase NDVI to rise above 0.42; optimize plant configuration and increase the number of trees planted
≥0.42	Guard against the excessive impact of new urban development on the current ecological level, reduce the excessive increase of hardened land surface, and avoid the decrease of NDVI to below 0.42

The impact of NDVI on heat island intensity is primarily evident in areas where the NDVI value exceeds 0.42, with a threshold of 0.42 for influencing heat island intensity. For areas with an NDVI less than 0.42, efforts should be made to encourage the development of three-dimensional greening, with a focus on integrating shrubs and trees to improve NDVI levels. In areas where the current NDVI is greater than 0.42, it is important to monitor and prevent excessive declines in ecological quality (Table 5.5).

5.1.2.5 Construction Land Regulation

The threshold level for the proportion of construction land affecting heat island intensity is 70%. When the proportion of construction land exceeds 70%, the probability of a management unit developing into a very strong or strong heat island area increases significantly. Since reducing the construction land area to alleviate heat island intensity is highly challenging, the focus should be on "prevention" rather than remediation. For regulatory planning units where the proportion of construction land has not yet reached 70%, a "warning line" should be established for areas in the suburbs with lower land values, implementing stricter approval standards for new construction. In contrast, in central areas where land values are higher, measures should be taken to mitigate the adverse effects of reducing new construction land on the thermal environment (Table 5.6).

5.1.2.6 Building Density Regulation

The impact of building density on heat island intensity is concentrated in areas where the building density is less than 0.1 and greater than 0.25, both showing a positive correlation, with a threshold level of 0.25 for the impact of building density on heat island intensity. The correlation analysis results in the previous text show that the correlation between building density and heat island intensity is the most significant

Table 5.6 Built-up land proportion control strategy

The current situation index range of the early warning object	Control method
<70%	Strictly control the increment and revitalize the stock, optimize the use efficiency of the existing construction land, and set up stricter approval standards for new construction land; the new construction land beyond the threshold value should further improve the green coverage requirements
≥70%	–

among all urban form indicators, so building density is a key control indicator for mitigating heat island intensity.

Considering the difficulty of demolishing existing buildings and the significant impact of building density on heat island intensity, urban form optimization targeting building density should emphasize both "prevention" and "management." For areas with a building density less than 0.25, which represent valuable low-intensity heat island or cold island zones (negative heat island intensity) in the city, future development should be approached with extreme caution. Building density increases in these sensitive areas should be strictly regulated through careful project approvals to prevent rapid growth. For areas where building density exceeds 0.25, urban form optimization should focus on old residential zones characterized by low-rise and multi-story buildings. In these areas, urban design principles should guide the demolition and reconstruction process to replace narrow-spaced low-rise buildings with wider-spaced high-rise structures, or to remove scattered low-rise buildings where demolition is less challenging (Table 5.7).

Table 5.7 Building density control strategy

The current situation index range of the early warning object	Control method
<0.25	Strictly control the increase of building density in the approval of new construction projects
≥0.25	On the premise of meeting the relevant urban design requirements, through the old city renovation, the overall demolition and reconstruction of residential areas that meet the current relevant technical specifications, encourage the replacement of low-rise buildings with high-rise buildings, or demolish low-rise scattered buildings that are less difficult to demolish, so as to reduce the building density

5.1.2.7 GSVF Regulation

The impact mechanism of GSVF on heat island intensity is complex. When the GSVF of the building combination is low, the mutual shading of buildings creates large shadow areas that mitigate heat island intensity. Conversely, when GSVF is high, the openness of the building combination promotes ventilation, making it challenging to achieve high heat island intensity. Statistically, the threshold interval of GSVF affecting heat island intensity is found to be 0.84–0.87, where the probability of forming extremely strong heat island areas and strong heat island areas exceeds 53%. Therefore, to optimize urban form and reduce heat island intensity using GSVF as a control indicator, the main focus should be on areas with a GSVF between 0.84 and 0.87. This can be achieved by replacing narrow-spaced low-rise buildings with wide-spaced high-rise structures through urban redevelopment, which reduces GSVF and increases the building shadow area. For areas with a current GSVF less than 0.84, where changing the existing high-density conditions is difficult, maintaining the status quo is recommended due to the large building shadow areas. For areas with a current GSVF greater than 0.87, new building projects should ensure that GSVF does not fall below 0.87, preserving openness and ensuring effective ventilation (Table 5.8).

5.2 Urban Form Guidance Technology for Different Types of Block

5.2.1 Block Type Division

From the mechanism analysis in the previous text, it is evident that the impact of each urban form indicator on heat island intensity varies across different areas. This variation implies that different areas require different urban form improvements to

Table 5.8 Ground surface view factor control strategy

The current situation index range of the early warning object	Control method
<0.84	Maintain the status quo pattern
0.84–0.87	Increase or decrease GSVF to make GSVF above 0.87 or below 0.84; demolish some scattered buildings to increase GSVF; increase the building height, reduce low buildings through old city reconstruction, and replace them with large-spacing high-rise buildings
>0.87	Ensure the exposure of the new building groups, so that their GSVF can not be lower than 0.87

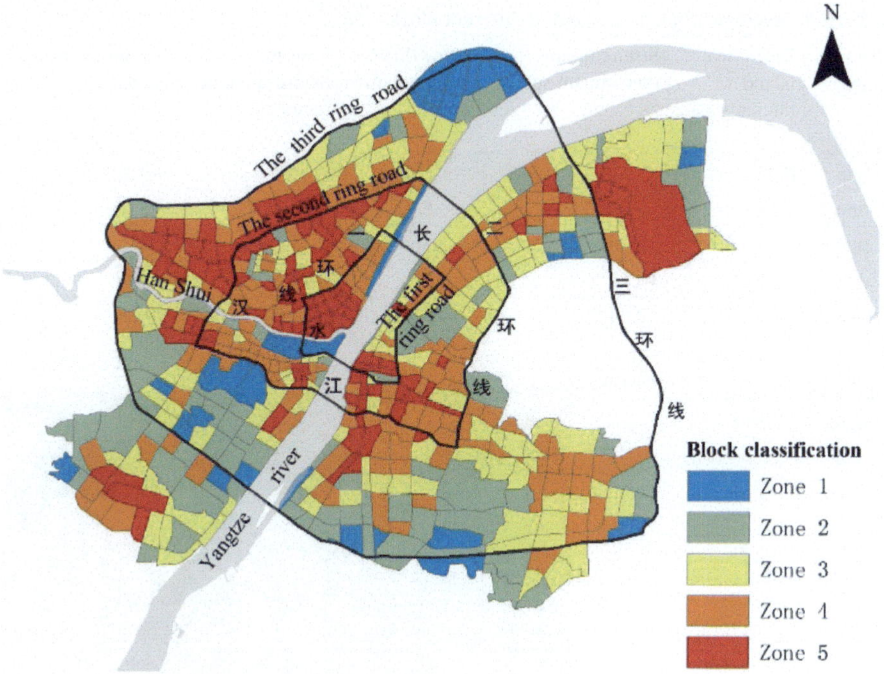

Fig. 5.11 Spatial layout of different type of block

effectively reduce heat island intensity. To achieve more targeted urban form control, distinct strategies should be employed for different areas based on the varying impacts of urban form indicators on heat island intensity. Consequently, this book classifies regulatory planning units into five categories based on the "Urban Climate Map" and the comprehensive mechanism analysis of 589 regulatory planning units (Fig. 5.11). Areas with higher comprehensive impact analysis scores experience greater thermal load from urban form, where the enhancement of the heat island effect is predominant. Conversely, areas with lower scores have stronger cooling capacities and are more effective in mitigating the heat island effect. Table 5.9 presents the scoring results for the enhancement and mitigation effects of urban form on heat island intensity across different zones.

5.2.1.1 Zone 1: Weak Thermal Load, Strong Cooling Type Blocks (Comprehensive Score: −14 to −7)

There are fewer blocks of this type, which are scattered around the periphery of the study area or in regions with a high proportion of vegetation and water bodies (Figs. 5.12, 5.17). Statistical analysis of urban form indicators reveals that these blocks have the highest NDVI, GSVF, green space rate, vegetation coverage, and soil

Table 5.9 Average urban heat island of different block type

Block type	Mean heat island intensity of (°C)	Strengthening mechanism score	Slow down mechanism score	External environmental factors score	Comprehensive score
Zone 1	0.92	1.61	−10.67	−0.17	−9.22
Zone 2	1.63	3.40	−8.58	1.20	−3.98
Zone 3	2.25	7.06	−7.36	1.38	1.08
Zone 4	2.80	10.11	−6.12	1.99	5.98
Zone 5	3.84	12.54	−4.52	2.84	10.86

Fig. 5.12 Typical spatial context of Zone 1

coverage, while exhibiting the lowest building density, proportion of industrial land, commercial land, construction land, and permanent population density. The scarcity of buildings in these blocks minimizes the potential for thermal loads from radiation heating and artificial heat dissipation. Additionally, the extensive permeable ground surfaces enhance evapotranspiration cooling from vegetation and provide excellent conditions for block ventilation, resulting in a heat island intensity of only 0.92 °C. In terms of external environmental influences, these blocks are significantly affected by heat diffusion from roads and concentrated industrial areas, while being less impacted by the cooling effect of large-scale water bodies.

5.2.1.2 Zone 2: Weak Thermal Load, Relatively Strong Cooling Type Blocks (Comprehensive Score: −6 to −2)

These blocks are predominantly located around the main water bodies within the study area and in regions with a high proportion of non-construction land, exhibiting a broader distribution than Zone 1. Similar to Zone 1, the urban form characteristics

Fig. 5.13 Typical spatial context of Zone 2

of these blocks include a higher proportion of permeable ground cover, lower proportions of various types of construction land, and reduced building density and permanent population density (Figs. 5.13, 5.17). With fewer artificial heat sources and a limited capacity for heat absorption through radiation due to the scarcity of buildings, the thermal load in these blocks is low, while ventilation is strong. Additionally, the proximity of these blocks to large-scale water bodies allows cold air generated above the water bodies to enter the blocks through air convection, effectively reducing the heat island intensity.

5.2.1.3 Zone 3: Relatively Strong Thermal Load, Relatively Strong Cooling Type Blocks (Comprehensive Score: −1 to 3)

Blocks of this type are distributed across different layers, typically situated on the periphery of densely populated urban areas. Urban form indicators for these blocks are moderate, with average construction intensity and proportion of construction land, as well as a general level of ecological ground cover (Figs. 5.14, 5.17). The thermal load and cooling effects are also moderate, resulting in an average heat island intensity. These blocks encompass urban fringe areas with low development levels and built-up areas with low construction intensity. Additionally, the influence of external environmental factors is minimal, with no significant positive or negative impact on the heat island intensity.

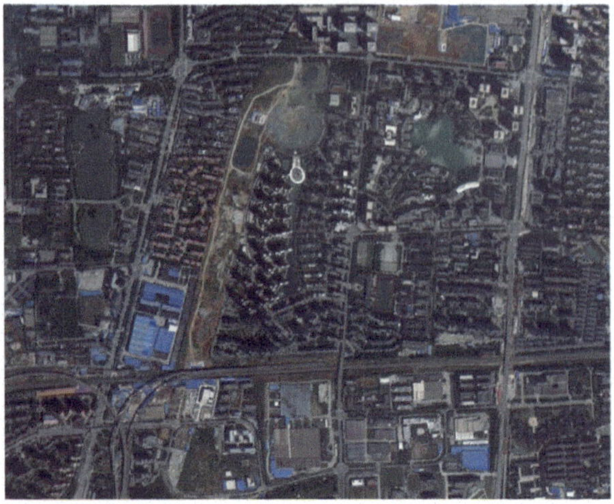

Fig. 5.14 Typical spatial context of Zone 3

5.2.1.4 Zone 4: Relatively Strong Thermal Load, Weak Cooling Type Blocks (Comprehensive Score: 4–8)

These types of blocks are concentrated within the second ring road, typically located in densely populated urban areas. Urban form indicators for these blocks show higher construction intensity and functional development indicators, but lower surface coverage indicators and GSVF (Figs. 5.15, 5.17), suggesting they generate significant anthropogenic and radiation heat. Due to the limited proportion of permeable surfaces, their heat dissipation capacity is poor, leading to the continuous accumulation of heat and a high heat island intensity. Additionally, the presence of high-density main roads in these densely populated urban areas further exacerbates the heat island intensity.

Fig. 5.15 Typical spatial context of Zone 4

5.2.1.5 Zone 5: High Heat Load, Weak Heat Dissipation Blocks (Composite Score: 9–16)

These types of blocks are concentrated within the first ring road, some areas along the second ring road, and industrial concentration areas around the third ring. From the perspective of morphological indicators, the proportions of industrial land, commercial land, permanent population density, building density, and construction land are all the highest compared to other types of blocks, while all types of surface coverage indicators and GSVF are the lowest (Figs. 5.16, 5.17), indicating that they have extremely strong radiation heat capacity, anthropogenic heat, and extremely weak ventilation and heat dissipation capacity. Among them, the areas within the first ring are greatly influenced by building density, permanent population density, and the proportion of commercial land, while some areas along the second ring road are influenced by building density and GSVF, and the peripheral industrial areas are mainly influenced by the proportion of industrial land. At the same time, due to the dense distribution of main roads in these areas or their proximity to concentrated industrial areas, they form a very strong heat island intensity under the combined effect of multiple factors.

5.2.2 The Correlation Between Urban Form Indicators and Heat Island Intensity in Different Types of Blocks

The correlation between various urban form indicators and heat island intensity exhibits spatial non-stationarity, meaning that the strength and direction of the correlation can vary across different areas, rather than maintaining a stable linear relationship everywhere [2–5]. Consequently, the regression coefficients and significance of the same urban form indicator on heat island intensity differ across areas, indicating

Fig. 5.16 Typical spatial context of Zone 5

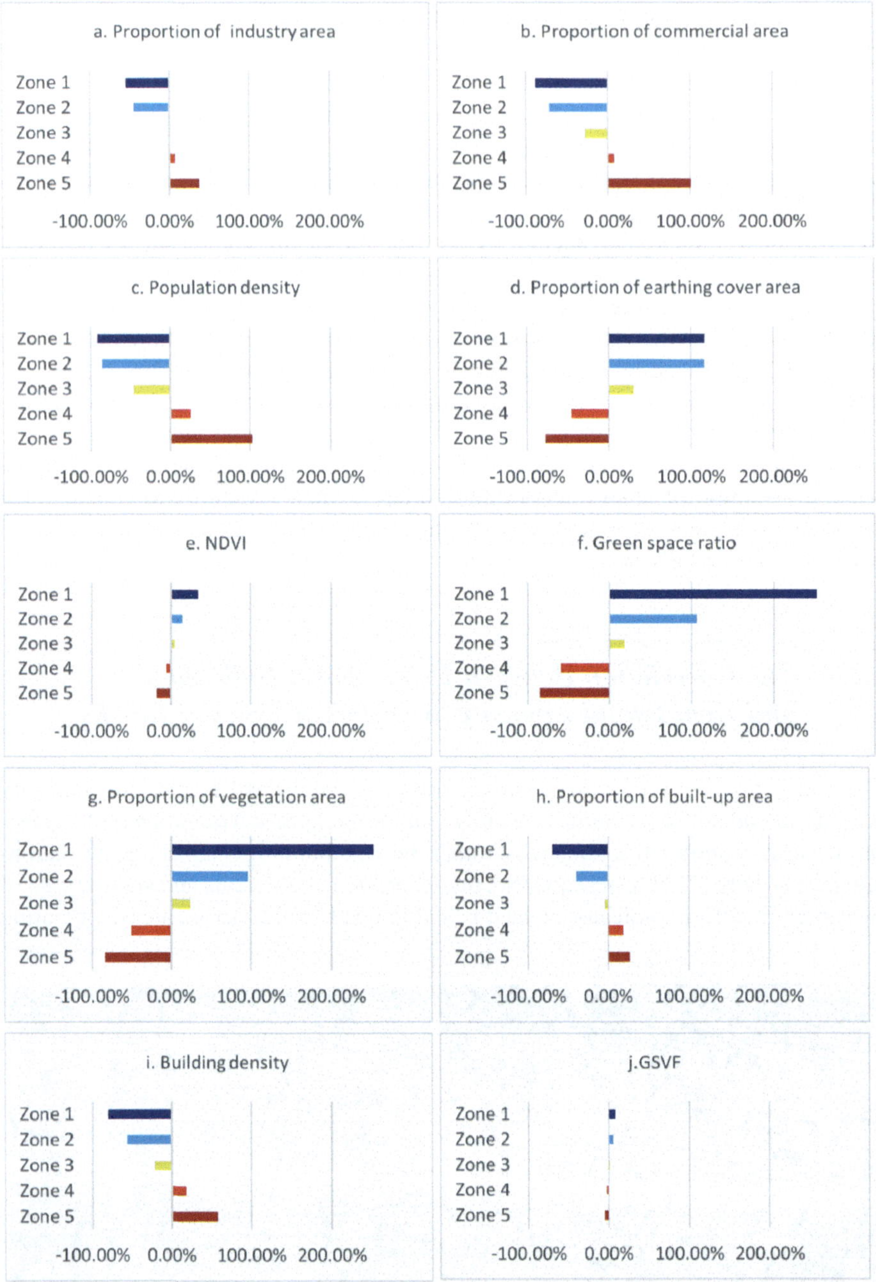

Fig. 5.17 Deviation between average urban form indexes and urban form indexes of different block type

that the mechanism by which urban form affects heat island intensity varies geographically. Therefore, this book will examine the differences in these correlations across different types of blocks.

To uncover the impact of urban form indicators on heat island intensity in each type of block, this book constructs a multiple linear regression model for each block type. The core urban form indicators serve as the explanatory variables, while the heat island intensity of each management unit acts as the dependent variable. The statistical results (Table 5.10) indicate that the correlation between urban form indicators and heat island intensity varies across different block types, and their explanatory power for heat island intensity differs accordingly. Generally, in blocks with high heat load and poor heat dissipation capacity, heat island intensity is strongly correlated with construction intensity, functional development, and building combinations. This is evidenced by significant correlations between urban form indicators and heat island intensity, with higher regression coefficients and a higher R^2 value, indicating good explanatory power. Conversely, in blocks with low heat load and good heat dissipation capacity, heat island intensity is more strongly correlated with surface coverage, resulting in a weaker explanatory power of urban form indicators, as manifested by a lower R^2 value.

Table 5.10 Multivariate linear regression standard coefficient of urban form indicators and urban heat island intensity in different types of blocks

	Zone 1	Zone 2	Zone 3	Zone 4	Zone 5
R^2	0.435	0.543	0.534	0.565	0.729
The proportion of industrial land (%)	0.081	0.028	0.123^{*}	0.259^{***}	0.263^{***}
The proportion of commercial land (%)	0.135	−0.011	−0.009	0.095^{*}	0.000
Population density (ten thousand people/km^2)	0.046	−0.135	−0.027	0.313^{***}	0.168^{*}
Ratio of green space (%)	0.360^{**}	0.110	−0.020	−0.087	0.098^{*}
The proportion of vegetation cover (%)	0.565	0.492^{**}	-0.251^{**}	0.034	−0.101
The proportion of soil cover (%)	0.563^{**}.	0.709	0.453^{***}	0.282^{***}	0.029
NDVI	−0.668	-0.406^{*}	0.137	-0.184^{**}	-0.222^{***}
Building density	0.187	0.371^{***}	0.420^{***}	0.345^{***}	0.526^{***}
The proportion of construction land (%)	0.179	0.210^{*}	−0.017	0.153^{**}	0.131^{**}
GSVF	−0.093	0.068^{***}	0.451^{***}	0.650^{***}	0.625^{***}

Note In parentheses are non-standard regression coefficients; *Note* ***, **, * means passing 1%, 5% and 10% significance level test respectively

5.2.2.1 Functional Development Indicators

The significant correlation between functional development indicators and heat island intensity is primarily concentrated in Zone 4 and Zone 5, both showing positive correlations. This indicates that functional development indicators strongly intensify heat island intensity in these block types, while in other block types, these indicators generally do not show significance. Among these, the strongest correlation is with permanent population density in Zone 4, with a regression coefficient of 0.313. Additionally, the proportion of industrial land in Zones 4 and 5 exhibits higher regression coefficients, at 0.259 and 0.263 respectively, whereas the correlation between the proportion of commercial land and heat island intensity is weaker.

5.2.2.2 Land Surface Coverage Indicators

In different types of blocks, certain surface coverage indicators show a significant correlation with heat island intensity, particularly in Zones 1, 2, and 3. NDVI and soil coverage are significant in these three types of blocks, while the proportion of vegetation coverage and green space rate show significance in the two block types. Notably, the strongest correlation is the proportion of soil coverage in Zone 1, with a regression coefficient of 0.563.

5.2.2.3 Construction Intensity Indicators

Building density shows a significant correlation with heat island intensity in all types of blocks except Zone 1. The regression coefficient reaches its maximum value in Zone 5, at 0.526. In Zones 2, 3, 4, and 5, building density has the largest standard regression coefficient among all core urban form indicators, indicating its substantial impact on heat island intensity. The proportion of construction land area is significant in Zones 2, 4, and 5, though the regression coefficient is not high.

5.2.2.4 Building Combination Indicators

GSVF shows a significant correlation with heat island intensity in all types of blocks except Zone 1. The regression coefficient is largest in Zone 4, reaching 0.650, with substantial values also observed in Zones 3 and 5. This indicates that the impact of GSVF on heat island intensity is primarily concentrated in blocks with high heat load and poor heat dissipation capacity.

5.2.3 Urban Form Regulation of Different Types of Blocks

Based on the correlation degree (with larger absolute regression coefficients indicating stronger correlations) between each urban form indicator and heat island intensity as shown in Table 5.11, the core urban form indicators are categorized into three groups: core regulation indicators (extremely strong correlation), key regulation indicators (strong correlation), and general regulation indicators (general correlation). Since the correlation between urban form indicators and heat island intensity varies across different block types, it is essential to adopt distinct urban form regulation strategies tailored to each type of block.

5.2.3.1 Zone 1: Blocks with Weak Heat Load and Strong Heat Dissipation

The core regulation indicator in this type of block is the proportion of soil coverage, and the key regulation indicator is the green space rate. The regulation strategies are as follows:

① Be alert to the blind expansion of built-up areas leading to a decrease in the green space rate, to avoid a rapid rise in heat island intensity.

② Focus on improving the level of vegetation coverage on the surface, building concentrated green spaces, and replacing existing soil-covered areas.

5.2.3.2 Zone 2: Low Heat Load, High Heat Dissipation Blocks

The key regulatory indicators in this type of block are vegetation coverage ratio, NDVI, construction land ratio, and building density, with GSVF as the general regulatory indicator. The regulatory strategies are as follows:

- Increase the ecological level of the ground surface, including increasing the vegetation coverage ratio and increasing the vegetation abundance to increase the NDVI through optimizing vertical greening.
- Be alert to the rapid increase in construction land, and strictly control the proportion of construction land in planning.
- Avoid rapid growth in building density, and increase the area of building shadows by regulating the form of building combinations.

5.2.3.3 Zone 3: High Heat Load, High Heat Dissipation Blocks

The key control indicators in this type of block are vegetation coverage ratio, soil coverage ratio, building density, GSVF, with industrial land ratio as the general control indicator. The regulatory strategies are as follows:

- This type of block has more soil coverage, which is one of the important factors to increase its heat island intensity. The ecological level of the soil-covered area should be improved, green spaces and forests should be planted on the existing soil, and the soil should be replaced with vegetation.
- Building density and GSVF are both directly proportional to the intensity of the heat island. The rapid growth of development and construction should be strictly controlled while encouraging the development of high-rise buildings in new areas to increase the area of building shadows and reduce radiant heat.
- Reduce the area of industrial land to avoid the adverse effects of heat generated by industry on residential areas.

5.2.3.4 Zone 4: High Heat Load, Low Heat Dissipation Blocks

The core control indicator in this type of block is GSVF, with industrial land ratio, permanent population density, soil coverage area, and building density as key control indicators, and NDVI, commercial land ratio, and construction land ratio as general control indicators. The regulatory strategies are as follows:

- Reduce the total building density by reducing industrial buildings.
- Increase the overall NDVI by constructing rooftop greening, wall greening, and enhancing the vegetation abundance of existing green spaces, thereby enhancing the evapotranspiration of moisture.
- Encourage the low-density development of high-rise buildings to increase the area of building shadows.

5.2.3.5 Zone 5: High Heat Load, Low Heat Dissipation Blocks

The core control indicators in this type of block are building density and GSVF, with industrial land ratio and NDVI as key regulatory indicators, and permanent population density, green space ratio, and construction land ratio as general regulatory indicators. The regulatory strategies are as follows:

- Reduce building density appropriately through urban renewal, while increasing building height, thereby increasing the area of building shadows through the staggered height of buildings.
- Increase the area and abundance of green spaces, thereby enhancing the evapotranspiration of moisture.
- Relocate industries embedded in residential areas to reduce the impact of industrial heat discharge on residential areas.

Based on the correlation between the urban form indicators in different partitions and the intensity of the heat island, this article proposes a planning strategy guide for reducing the heat island intensity in different partitions (Table 5.11).

Table 5.11 Planning strategies for reducing urban heat island intensity in different zones

<table>
<tr><th>Block type</th><th colspan="2">Distribution area</th><th colspan="2">Block characteristics</th><th>Planning countermeasures</th></tr>
<tr><td rowspan="4">Zone 1</td><td colspan="2">Distribution in the periphery of the built-up area; closely combined with large-scale water</td><td colspan="2">The heat island intensity is very low;
It is usually an undeveloped area with few buildings, little construction land, and large-scale vegetation or water cover
The thermal load is extremely low and the heat dissipation capacity is extremely strong</td><td>The blind expansion of built-up area leads to the decline of green land rate
Focus on improving the vegetation coverage level of land cover, building large-scale green space, and replacing the existing land cover area</td></tr>
<tr><td colspan="5">Main control indicators*</td></tr>
<tr><td colspan="2">Ratio of green space</td><td colspan="3">The proportion of soil cover</td></tr>
<tr><td colspan="2">☆☆</td><td colspan="3">☆☆☆</td></tr>
<tr><td rowspan="4">Zone 2</td><td colspan="2">Centralized distribution in the periphery of the built-up area; mainly distributed outside the second ring; and closely combined with large-scale water body</td><td colspan="2">The heat island intensity is low;
There are few industrial and commercial functions, few buildings, and usually undeveloped areas with a relatively general level of vegetation cover;
Low thermal load, high heat dissipation capacity</td><td>Increase the ecological level of the surface;
Be alert to the rapid increase in construction land;
Avoid the rapid growth of building density, and increase the shadow area of buildings by regulating the form of building combinations</td></tr>
<tr><td colspan="5">Main control indicators*</td></tr>
<tr><td>The proportion of vegetation cover</td><td>NDVI</td><td>The proportion of construction land</td><td>Building density</td><td>GSVF</td></tr>
<tr><td>☆☆</td><td>☆☆</td><td>☆☆</td><td>☆☆</td><td>☆</td></tr>
</table>

(continued)

Table 5.11 (continued)

Block type	Distribution area	Block characteristics	Planning countermeasures
Zone 3	Centralized distribution on the edge of concentrated built-up area; distributed in all circles	Heat island strength is moderate; Usually not fully developed area, the urban function is not sound, but the construction intensity is general Moderate thermal load and heat dissipation capacity	To improve the ecological level of the land cover area, planting green space and forest land on the current land cover, replacing the land cover with vegetation; Strictly control the excessive growth of development and construction, encourage the development of high-rise buildings in new areas, increase the shadow area of buildings, and reduce the radiation heat gain Reduce the area of industrial land

Main control indicators*

The proportion of industrial land	The proportion of vegetation cover	The proportion of soil cover	Building density	GSVF
☆	☆☆	☆☆	☆☆	☆☆

(continued)

Table 5.11 (continued)

Block type	Distribution area	Block characteristics	Planning countermeasures
Zone 4	Located in the densely-built area; mainly distributed in the second ring Road	The heat island intensity is strong; It is usually a densely built-up area with sound urban functions, a certain amount of industrial and commercial land, and high construction intensity The heat load is high and the heat dissipation capacity is weak	Reducing total building density by reducing industrial construction; The overall NDVI was improved by constructing roof greening, wall greening and increasing the vegetation abundance of existing green space Encourage low-density development of high-rise buildings and increase the shadow area of buildings

Main control indicators*							
The Proportion of industrial land	The Proportion of commercial land	Population Density	The Proportion of soil cover	NDVI	Building density	The proportion of construction land	GSVF
☆☆	☆	☆☆	☆☆	☆	☆☆	☆	☆☆☆

(continued)

Table 5.11 (continued)

Block type	Distribution area	Block characteristics	Planning countermeasures
Zone 5	Centralized distribution in the first ring line, part of the area along the second ring road and centralized industrial zone	It is usually a well-developed old city or industrial area with a high density of buildings; Enrichment of industrial and commercial functions; Basically no large open space; The ecological level of the surface is very poor, basically covered by the hardened surface	Through the reconstruction of the old city, the building density is reduced appropriately, while the building height is increased, so as to increase the shadow area of the building through the staggered height of the building Increase green area and vegetation abundance; The industry embedded in residential areas will be relocated to reduce the impact of industrial heat exhaust on residents' living areas

Main control indicators*						
The proportion of industrial land	Population density	Ratio of green space	NDVI	Building density	The Proportion of construction land	GSVF
☆☆	☆	☆	☆☆	☆☆☆	☆	☆☆☆

☆☆☆: Core regulation index; ☆☆: key regulation index; ☆: general regulation index

* The main control indicators of each block type are derived from "6.2.2 Correlation between spatial form indicators and heat island intensity in different types of blocks", that is, indicators that have a significant correlation with heat island intensity. Among them, in the multiple linear regression model, the index with the standard regression coefficient greater than 0.5 is the "core regulation index", the index with the standard regression coefficient 0.2–0.5 is the "key regulation index", and the index with the standard regression coefficient less than 0.2 is the "general regulation index"

5.3 Urban Form Optimization Techniques for Special Function Zones

5.3.1 Avoid the "Overlapping Effect" Between High-Temperature Zones, Increase Green Coverage Along Main Roads and Concentrated Industrial Areas

Research findings indicate that main roads and concentrated industrial areas are significant spatial elements contributing to high-intensity heat island zones (see Sect. 5.4.1). These elements facilitate the formation of extensive, continuous high-temperature zones through heat transfer. The interaction between these high-temperature zones via convective heat transfer leads to an "overlapping effect" where high-temperature areas merge. To mitigate the adverse impacts of extreme heat island intensity on surrounding residents, it is essential to enforce higher green space rates and open space requirements for areas along major roads and concentrated industrial zones. Furthermore, the extensive high-temperature zones created by industrial areas obstruct air convection between densely populated inner areas and the outer suburbs, hindering the inflow of cool suburban air. To counteract this, it is crucial to enhance vegetation, lower building density, and establish ventilation corridors to facilitate airflow between the suburbs and residential areas.

5.3.2 Extend the Cooling Area of Water Bodies, Reduce the Construction Intensity of Waterfront Areas

The spatial characteristics of heat islands, such as "gradient buffering" and the varying intensity in weaker heat island areas, demonstrate that low-temperature zones dominated by water bodies can alleviate heat island intensity in adjacent areas. However, in Wuhan's main urban area, major lakes and the Yangtze River are currently obstructed by high-intensity construction zones, which impede air circulation between these water bodies and their surroundings. Despite Wuhan's abundant water resources, the cooling effect of these bodies of water is not fully realized due to this obstruction.

To maximize the cooling effect of large-scale water bodies, development and construction in their surrounding areas should be carefully regulated, emphasizing a well-considered spatial layout. Firstly, land use planning around these water bodies should incorporate ventilation corridors aligned with the city's prevailing wind direction. This approach aims to channel cooler air from the water body into the downwind residential areas and to direct heat from upwind areas toward the water for cooling. Secondly, increasing the porosity of building arrangements in the vicinity of water bodies is crucial. This can be achieved by avoiding the construction of large, slab-like buildings facing the water, thereby reducing airflow resistance between the water body and its surroundings.

5.4 Case Study on Reducing Heat Island Intensity—Taking Hualou Street Block as an Example

5.4.1 Analysis of Current Urban Form of Hualou Street Block and Control Indicators Identification

The Hualou Street neighborhood is a historic district situated around Hualou Street, adjacent to Han Zheng Street in Wuhan City. It is one of the oldest neighborhoods in Hankou, preserved for an extended period, and is part of the Minquan Road area, which is identified as one of the typical extremely high heat island zones discussed before. Predominantly residential, the neighborhood features numerous streets, including Minzu Road, Tongyi Street, Hualou Street, and Ersheng Lane. These streets experience heavy pedestrian traffic and are characterized by a mix of residential and commercial properties, reflecting the old city's vibrant atmosphere (Fig. 5.18).

Commercial facilities are densely located along the streets, consisting mainly of storefronts that offer daily necessities, small restaurants, and leisure and entertainment options, with a notable absence of independent commercial land. The area is well-equipped with public service facilities such as Daxing Road Primary School, Daxing Road Kindergarten, and Hankou Hui People's Primary School. The current topographic map reveals that akin to other parts of old Hankou, the neighborhood is marked by small, low-rise buildings that are closely spaced, resulting in a densely packed architectural environment (Fig. 5.19).

Statistical results indicate that several urban form indicators for the Hualou Street neighborhood exceed the warning thresholds for extremely strong and strong heat island areas, including permanent population density, NDVI, vegetation coverage ratio, building density, and GSVF. These indicators suggest a high risk of forming high-intensity heat island areas (Table 5.12).

Firstly, the building density in this area is 0.60, significantly surpassing the maximum density limits set by the "Urban Residential Area Planning and Design Specifications" (low-rise: 0.4; multi-story: 0.3; mid-high-rise: 0.28; high-rise: 0.20). High building density obstructs urban ventilation and prevents the establishment of effective ventilation corridors, leading to concentrated human activity and heat discharge, which exacerbates the urban thermal environment.

Secondly, the area's GSVF value is 0.85, with an average building height of 3–4 stories. This combination contributes to high heat island intensity as it impedes large-scale shading and allows direct solar radiation on hardened surfaces and

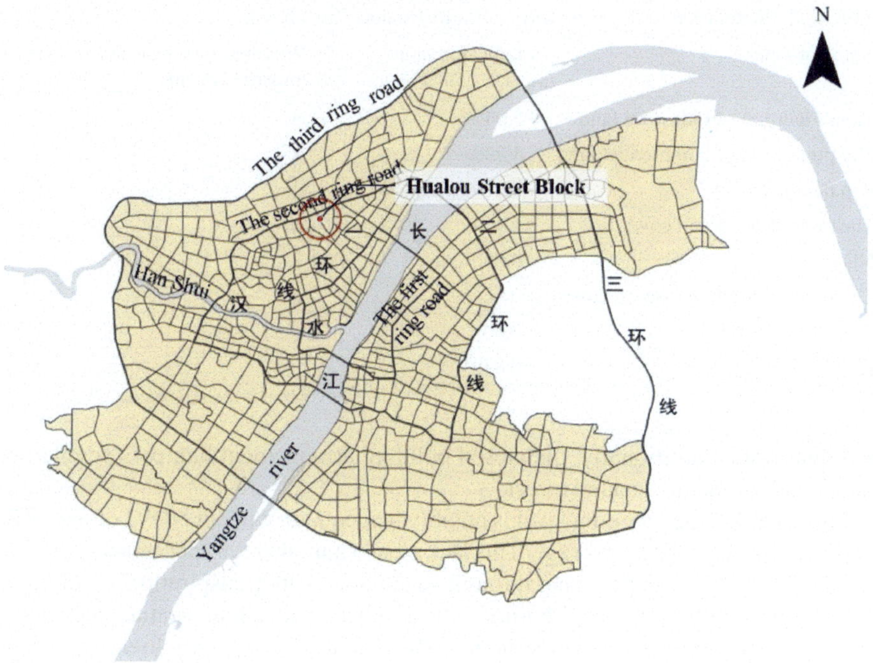

Fig. 5.18 Location of Hualou district

Fig. 5.19 Buildings of Hualou district

Table 5.12 Statistics of core urban form index of Hualou Street block

Name of index	Indicator value	Whether it exceeds the warning threshold or not
The proportion of industrial land	5.04%	×
The proportion of commercial land	4.36%	×
Population density	83,700 people / km^2	√
The proportion of soil cover	1.45%	×
NDVI	0.06	×
The proportion of vegetation cover	0	√
Building density	0.60	√
GSVF	0.849	√

building roofs. Additionally, the closed building arrangement hampers ventilation, compounding the heat island effect.

Furthermore, the greening level is notably inadequate, with an NDVI of only 0.06 and minimal vegetation coverage. This lack of substantial green space means the area cannot effectively dissipate heat through vegetation evapotranspiration, resulting in heat retention within the neighborhood. In summary, the unique spatial characteristics of the Hualou Street neighborhood contribute to its extremely high heat island intensity.

According to the urban form differentiation guidelines for various neighborhood types, the Hualou Street neighborhood falls within Zone 5. In this zone, the most influential indicators affecting heat island intensity are building density and GSVF. The conclusions from the urban form indicator warning system reveal that the building density and GSVF in this neighborhood significantly increase the likelihood of forming strong and extremely strong heat island areas and have a substantial impact on heat island intensity.

Therefore, this chapter emphasizes urban form control strategies for the Hualou Street neighborhood, focusing specifically on building density and GSVF. Targeted control strategies for building layout are proposed to address these two critical indicators.

5.4.2 *Thermal Environment of Hualou Street Neighborhood*

Thermal environment simulations for the Hualou Street district were conducted using Pheonics software. The simulation parameters were set as follows: wind direction was set to 22.5° east of south, wind speed at 1.5 m above the ground was set to 4 m/s, and the solar angle was set to the instantaneous angle at 10:55 on October 6, 2014. Direct solar radiation was set to 500 W/m^2, and diffuse radiation was set to 250 W/m^2. The grid division for each simulation case was consistent, with 700 units evenly distributed in the north–south direction, 600 units in the east–west direction, and 20 units in the vertical direction, with further subdivisions within the block. Each case assumed a flat ground surface.

The architectural space combination was modeled in SketchUp software based on Wuhan city architectural data. The ground surface emissivity was calculated using the NDVI value for each case, and the solar absorption index was uniformly set to 0.65. Building emissivity and solar absorption index values were set to 0.9862 and 0.72, respectively, according to relevant design plans. The heat transfer coefficient for both the underlying surface and buildings was set to 10 $W/m^2/°C$. Initial conditions included a building surface material temperature and environmental temperature of 20 °C, and an initial ground temperature of 15 °C.

After setting these conditions, the radiation model was loaded, and the spatial grid was refined according to the case scale. The thermal environment simulation results were obtained through 1000 iterations, providing the spatial distribution of wind speed and temperature at the 2.0-m horizontal plane within each case under identical external conditions.

The simulation results indicate that higher temperatures are observed in areas with extremely small building spacings and high building densities, primarily concentrated around the buildings. In contrast, temperatures in open spaces are generally lower (Fig. 5.20). Possible explanations for this observation include: 1) In densely built-up areas, buildings receive more solar radiation, heating the surrounding air. With reduced wind speed, heat dissipation is limited, resulting in higher temperatures. 2) Solar radiation distribution analysis shows that some high-rise buildings create significant shadow areas, preventing direct solar radiation from reaching the ground. Although wind speed may be reduced in these shadowed areas, the overall heat generation is lower, leading to lower temperatures.

5.4.3 Comparison of Thermal Environment in Multiple Renovation Modes

5.4.3.1 Renovation Mode Design

The building density in the Hualou Street district significantly exceeds the threshold for forming strong heat island areas. Previous analyses have indicated that building density has a substantial impact on heat island intensity. Consequently, the renovation plan prioritizes reducing building density as a key strategy for optimizing the thermal environment in this area. This book explores various renovation plans to assess their effects on the overall thermal environment of Hualou Street, with a focus on reducing building density while maintaining equal levels of density reduction across different scenarios.

This book proposes four renovation modes:

- Courtyard Renovation Mode
- Scattered Point Renovation Mode

Fig. 5.20 Air temperature simulation of Hualou district

- Corridor Renovation Mode
- Wedge Mouth Renovation Mode

All four renovation modes aim to reduce the building density from the current 0.60–0.55. The Courtyard and Scattered Point modes focus on demolishing buildings within the block, with the distinction that the Courtyard mode involves contiguous demolition, while the Scattered Point mode involves non-contiguous demolition. Conversely, the Corridor and Wedge Mouth modes address demolition both inside and outside the block. The Corridor mode targets block-edge and strip-like demolitions within the block, while the Wedge Mouth mode focuses on demolishing buildings in block-distributed and strip patterns throughout the block.

The proposed renovation plans are detailed as follows:

- Courtyard-Type Renovation Model. This model seeks to enhance internal ventilation by creating more open space within the block. It involves demolishing a cluster of buildings to form a hollow environment, thereby improving air circulation (Fig. 5.15a).
- Scattered Renovation Model. This model aims to enhance the thermal environment by selectively demolishing buildings in the most densely populated and poorly ventilated areas within the block (Fig. 5.15b).
- Corridor-Type Renovation Model. This model intends to facilitate external wind flow into the block by demolishing a row of buildings aligned with the prevailing wind direction, thereby reducing internal heat accumulation (Fig. 5.15c).
- Wedge-Mouth Renovation Model. This model creates three wedge-shaped open spaces on the east, south, and north sides of the block through targeted demolition.

The objective is to use these open spaces to channel and accumulate wind flow, enhancing overall ventilation (Fig. 5.15d).

5.4.3.2 Comparison of Renovation Models with the Thermal Environment of the Current Block

The external condition parameters used in the current thermal environment analysis were applied to simulate each renovation plan. The simulation results indicate varying impacts on temperature for each plan.

Courtyard-Type Renovation Mode resulted in a slight temperature reduction in a small area. However, the temperature in most of the newly created open spaces and other areas remained unchanged. This may be due to the plan not significantly increasing wind speed within the open spaces, thus failing to produce a noticeable cooling effect (Fig. 5.21a).

Scattered Renovation Model significantly reduced temperatures in local areas by creating small-scale open spaces, particularly in narrow streets and alleys where building spacing is extremely tight. The improvement in the heat island effect is likely because, with the removal of some buildings, the air in these narrow spaces is only heated by the surface of one side of the remaining buildings, leading to a temperature drop (Fig. 5.21b).

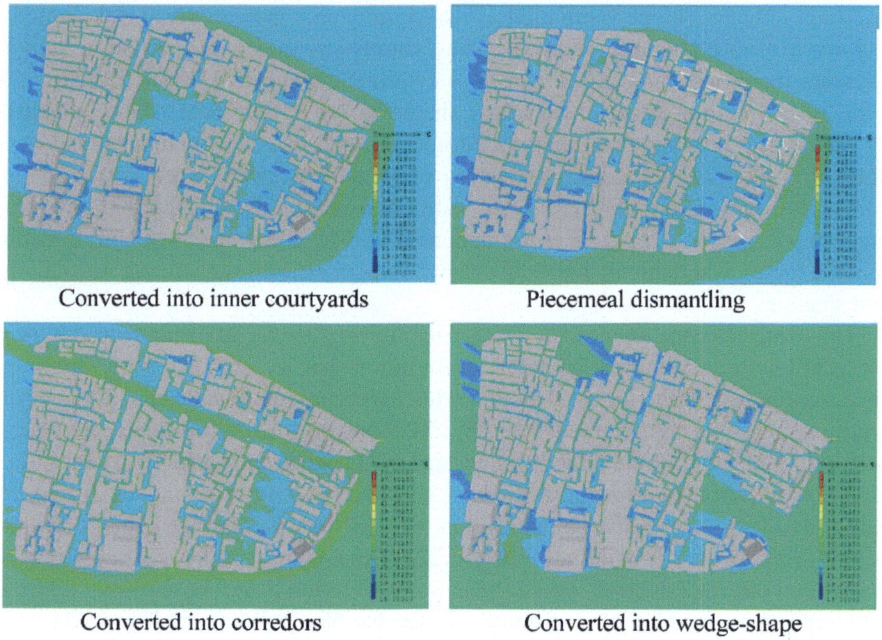

Fig. 5.21 Thermal environment analysis of blocks in different transform modes

Corridor-Type Renovation Model created a high-temperature corridor, where the temperature in the center of the corridor increased significantly while the temperatures on both sides were lower, forming a streamlined shape. The likely reason is that the corridor's central ground, previously shaded by buildings, is now exposed to direct sunlight, heating the air above it. The wind flow through the corridor further contributes to this streamlined high-temperature zone (Fig. 5.21c).

Wedge-Mouth Renovation Model led to a significant temperature increase in the southeast corner, especially in the alley downwind of the wedge mouth. The demolition of mid- and high-rise buildings in this area reduced building shadows, allowing direct sunlight to heat the ground and raise the temperature. Additionally, the opening direction of the southeast wedge mouth aligns with the main wind direction, allowing high-temperature air to enter the block's interior (Fig. 5.21d).

5.4.4 Comprehensive Optimization Plan Design and Thermal Environment Analysis

The comparative analysis of the above schemes reveals that the block's thermal environment is influenced by both ventilation convection and solar radiation. To mitigate the heat island effect, suitable architectural combination optimization schemes are necessary. In the corridor-type and wedge-type renovation modes, although reducing buildings enhances block ventilation, it also significantly increases heat island intensity due to reduced building shadows. Thus, simply strengthening ventilation cannot alleviate the heat island effect and may even exacerbate it. High-rise buildings, with lower building density and larger shadows, can reduce heat island intensity by improving ventilation and shading. Moreover, high-rise construction can balance demolition, aiding in funding renovation projects. Therefore, a renovation mode combining demolition and high-rise construction is effective for thermal environment improvement and practical implementation. The comprehensive optimization scheme focuses on reducing building density and GSVF while enhancing ventilation and shading.

The comprehensive optimization scheme proposed in this article aims to enhance the ventilation environment by reducing building density and increasing building shadows through high-rise construction while adhering to urban design and control requirements. This improvement plan (Fig. 5.22) balances demolition and reconstruction without changing the total building area of the existing block. The demolition plan is based on the comparative analysis of different renovation modes discussed earlier and combines scattered demolition, wedge-shaped demolition, and corridor demolition (see the red buildings in Fig. 5.22). The goal is to improve air convection within and outside the block, while the new construction plan focuses on increasing building shadows with high-rise buildings (see the blue buildings in Fig. 5.22).

The specific optimization plan includes the following steps:

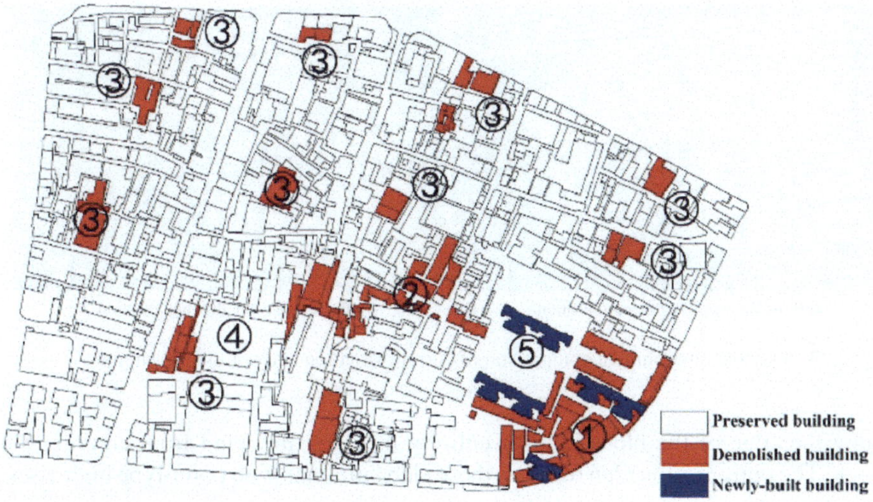

Fig. 5.22 Optimized spatial layout of buildings

- **Wedge-Shaped Wind Storage Space**:
  Demolish buildings in the southeast upwind direction to create a wedge-shaped wind storage space, connecting internal open spaces with the block's exterior (see Fig. 5.22a, ①).
- **Ventilation Corridor**:
  Demolish buildings downwind of the wedge-shaped space to create a ventilation corridor, linking the wedge-shaped space with another internal vacant lot (see Fig. 5.22a, ②, ④).
- **Scattered Demolition**:
  Implement scattered demolition in areas with high local building density, forming small-scale vacant lots (see Fig. 5.22a, ③).
- **High-Rise Construction**:
  Construct a few point-type high-rises (18–21 floors) in the southeast corner. These new buildings, equal in area to the demolished ones, are oriented perpendicularly to the wind direction to avoid wind obstruction and create large building shadows (see Fig. 5.22a, ⑤).

Compared to the current plan, the comprehensive optimization plan reduces building density while increasing average building height, lowering the GSVF value, and maintaining the total building area. This approach effectively balances "reducing density and increasing height." (Fig. 5.23)/

The comprehensive optimization plan was subjected to the same boundary conditions as the previous thermal environment simulations (Fig. 5.22). The results indicate that compared to the current thermal environment, the temperature in some areas has decreased. In the current layout, slab buildings in the southeast corner are positioned

Fig. 5.23 Thermal environment comparison of current situation and optimized scheme

along the edge of the block, almost entirely blocking the wind from entering the block. The optimization plan replaces these slab buildings with point-type high-rises, allowing external wind to penetrate the block.

This change has significantly increased wind speed in the open space in the southeast corner, with the accumulated wind energy spreading westward into the newly created ventilation corridor and further into the interior of the block. Additionally, high-rise buildings create extensive shadows on the ground, reducing heat radiation and cooling the wind storage node in the upwind area. The external wind is thus "cooled" before entering the block.

Despite a reduction in building density, the comprehensive optimization plan avoids the simultaneous increase in wind speed and temperature observed in the corridor-type renovation plan. In the corridor-type plan, although building density is reduced, the GSVF also decreases, leading to increased exposure of the ground to solar radiation and forming a high-temperature corridor. In contrast, the comprehensive optimization plan not only reduces building density but also lowers the GSVF through high-rise buildings, reducing the temperature in the upwind direction. Additionally, by adopting a scattered demolition strategy, the plan further reduces building density in densely built areas, suppressing the large-scale radiation heat of buildings and effectively mitigating the high-intensity heat island effect in local areas.

5.4.5 Urban Form Optimization Strategy

5.4.5.1 Construct Wind Storage Nodes and Ventilation Corridors to Reduce Building Combination Heat Accumulation

When reducing building density, the wedge-type and corridor-type renovation modes significantly increase wind speed compared to the courtyard-type and scattered-type modes. This indicates that constructing wind storage nodes in the upwind direction and building ventilation corridors along the main wind direction can optimize the

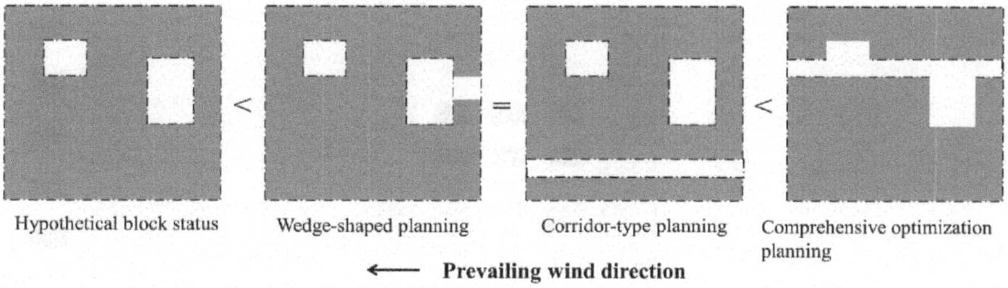

Fig. 5.24 Schematic diagram of "wedge + corridor" control strategy

block's wind environment, enhancing convective heat dissipation and reducing heat accumulation from building clusters. Therefore, the comprehensive optimization plan integrates the advantages of both approaches, creating a "wedge + corridor" design to enhance ventilation (Fig. 5.24).

The current building layout of the Hualou Street block obstructs the open space in the upwind direction with slab buildings, turning it into a calm wind area and wasting its ventilation potential. Additionally, this open space is densely surrounded by buildings, preventing accumulated wind energy from spreading into the block. The narrow internal streets also hinder the introduction of external wind, creating many calm wind areas that trap heat.

On this account, to improve ventilation convection and reduce heat accumulation, the optimization plan should consider the unique building configuration of the current block. Priority should be given to demolishing buildings in the upwind direction where feasible, and constructing high-rise buildings with larger spacing to form an open space that can accumulate wind energy. This open space should guide external wind into the block, which can then be further directed into the block through a ventilation corridor closely connected to the wind storage node.

5.4.5.2 Applying New High-Rise Buildings to Increase Building Shadows and Reduce Heat

Although the wedge-shaped and corridor renovation models significantly increase wind speed within the block, their cooling effects are poor. Areas with high wind speed, such as the wedge-shaped area and the central area of the ventilation corridor, actually exhibit higher temperatures. This is primarily due to the reduction in building shadows caused by demolition, which leads to increased surface radiation heat. The presence of building shadows is crucial for the overall thermal environment. By strategically using the shadow areas created by a combination of ventilation spaces and buildings, we can create wedge-shaped areas and ventilation corridors covered by building shadows. This approach improves the wind environment and lowers temperatures in the upwind direction, helping to gather low-temperature air at the

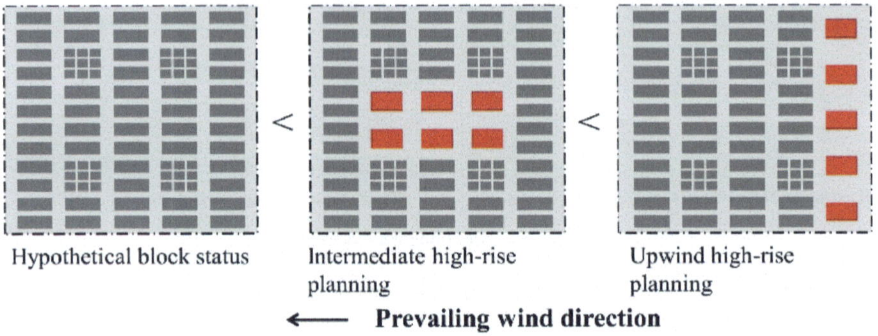

Fig. 5.25 Strategy of increasing high-rise buildings

wedge and inject it into the block along the dominant wind direction, thereby reducing the heat island intensity.

High-rise buildings can serve as reconstruction areas for demolished buildings, allowing demolition measures and high-rise construction to be carried out simultaneously, enhancing the plan's operability. The comprehensive optimization plan addresses the drawback of simple demolition plans, which reduce building shadows and consequently raise temperatures. Therefore, several point-type high-rise buildings are constructed at wind storage nodes located in the upwind direction, forming extensive building shadows in the open space without obstructing the dominant wind direction (Fig. 5.25). The simulation results show that this plan cools the external wind passing through these areas and avoids the temperature rise caused by ground surface exposure as seen in the wedge-shaped demolition plan. This demonstrates that constructing wind storage nodes and ventilation corridors must be coordinated with building shadows. Increasing building shadows through high-rise buildings in the upwind direction can reduce the temperature of the transit wind.

5.4.5.3 Scattered Reduce Building Footprint and Reduce Artificial Surface Heat Transfer

The scatter-type demolition method, which reduces building density and artificial surface heat exchange, is also effective. Simulation results show that among the four demolition plans, the scatter-type renovation model, although unable to spread its mitigation effect on heat island intensity across the entire block, significantly alleviates high temperatures in local areas. Therefore, the comprehensive optimization plan adopts scatter-type demolition in areas that are difficult to radiate to the wedge and ventilation corridor, demolishing some scattered buildings in areas with extremely high building density (Fig. 5.26). Simulation results indicate that temperatures in the surrounding areas drop significantly after demolition. The principle behind scatter-type demolition is that it reduces the building surface area available for radiation heating while maintaining small continuous demolition volumes. The

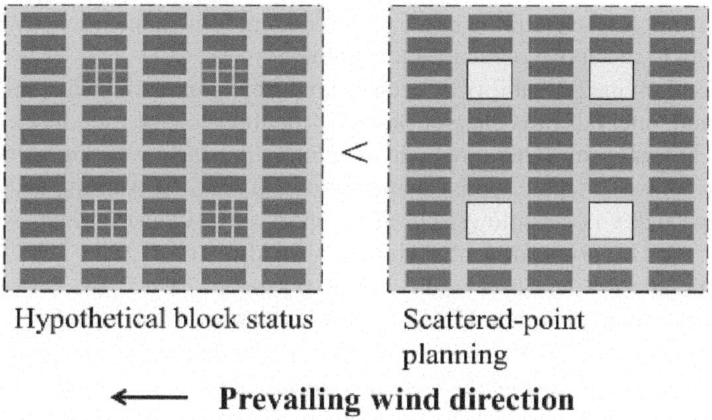

Fig. 5.26 Schematic diagram of "dotted remove" control strategy

resulting courtyard spaces are still enclosed by buildings, preventing direct sunlight from hitting the ground and thereby reducing heat island intensity. When promoting this method on a larger scale, it is essential to target small-volume scattered buildings for demolition and avoid reducing building shadows.

In conclusion, this chapter proposes three urban form index control technologies: "urban form index warning and optimization technology," "differential guidance techniques for urban form of different types of blocks," and "optimization techniques for urban form of special function areas." The first focuses on specific urban form optimization techniques within a block, while the latter two distinguish between different types of blocks and discuss the varying regulatory focuses for different zones. These urban form control techniques provide operable planning strategies to reduce heat island intensity in different urban areas. Finally, the chapter applies these control techniques to the Hualou Street block in Wuhan as a case study for reducing heat island intensity through planning practices.

5.4.6 Chapter Conclusion

This chapter puts forward three urban form index control technologies, namely "urban form index early warning and optimization technology", "urban form differentiation guidance technology for different types of blocks" and "urban form optimization technology for special functional areas". The former focuses on discussing which specific urban form optimization technologies should be adopted inside specific blocks. The latter two distinguish different types of blocks and discuss the differences in regulatory emphasis in different districts. The above three urban form control techniques can provide operational planning countermeasures for different urban areas to reduce heat island intensity. Finally, taking Hualou Street in Wuhan

City as an example, the planning practice of reducing heat island intensity is carried out according to the above regulation technology.

First, based on the probability of each index forming a strong heat island area in different value intervals, this chapter proposes the threshold of each index affecting the heat island intensity. This threshold is actually the early warning value of heat island intensity based on the spatial shape index, and the early warning value of each spatial shape index is as follows: (1) the proportion of industrial land is "greater than 24%"; ② The proportion of commercial land is "more than 10%"; (3) The permanent population density is "more than 30,000 people /km2"; (4) The proportion of vegetation cover is "less than 15%"; ⑤ NDVI is "less than 0.42"; ⑥ The proportion of construction land is "greater than 70%"; ⑦ Building density is "greater than 0.25"; ⑧ GSVF is "between 0.84 and 0.87". When the urban form index exceeds the range of warning value, it is easy to form an extremely strong heat island area and a strong heat island area. Therefore, the planning and regulation of urban form should focus on whether future urban development will cause some core urban form indicators to exceed the threshold range, and the occurrence of this phenomenon should be avoided or delayed as much as possible through urban form regulation.

Second, Chapter 5 reveals that while the influence mechanisms of urban form indicators on heat island intensity remain consistent, their effects vary across different regulatory planning units due to spatial heterogeneity. Specifically, in regulatory planning units with lower comprehensive mechanism scores, urban form indicators have a smaller impact on intensifying heat islands and a larger effect on mitigating them. Conversely, in units with higher scores, the impact of urban form indicators on heat island intensity is more pronounced. To address these differences, this chapter proposes targeted urban form control strategies based on comprehensive mechanism scores. For areas with lower scores, the focus should be on preserving existing ecological conditions and monitoring any rapid increases in construction intensity or building combination indicators resulting from urban development. In contrast, for areas with higher scores, strategies should include enhancing vegetation coverage, promoting three-dimensional greening, and adopting development modes that either emphasize high-density, low-height structures or low-density, high-height structures. Additionally, increasing building shadow areas can help reduce the absorption of solar radiation by hardened surfaces.

Third, based on the impact of large-scale water bodies and concentrated industrial areas on heat island intensity, the urban form of special function areas can be optimized by controlling urban form indicators. This optimization aims to mitigate the adverse effects of high temperatures on residents.

Finally, to implement index control requirements at the urban planning and design level, this chapter uses the Hualou Street block in Wuhan as a case study. Based on the block division type identified in Sect. 6.2.1, Hualou Street is classified as Zone 5—characterized by strong heat load and weak heat dissipation. Consequently, the key control indicators for this block are identified as building density and Ground Surface Ventilation Factor (GSVF), as discussed in Sect. 6.2.3. The calculation of core urban form indicators reveals that the building density index significantly exceeds the threshold range, making building density and GSVF the primary focus for optimization. To address this, four transformation modes are proposed: ① Courtyard Transformation Mode, ② Scattered Point Transformation Mode, ③ Corridor Transformation Mode, and ④ Wedge Mouth Transformation Mode. After evaluating the thermal environment under these modes, a comprehensive optimization scheme is developed, encompassing three main strategies:

- Construction of Wind Storage Nodes and Ventilation Corridors: This strategy aims to reduce heat accumulation from building combinations.
- Increasing High-Rise Buildings in the Upwind Direction: This approach focuses on reducing heat by enhancing and managing building shadows.
- Scattered Rectification of Building Density: This strategy seeks to lower artificial surface heat exchange through selective demolition.

These strategies collectively aim to effectively reduce heat island intensity and improve the thermal environment of the Hualou Street block.

References

1. Yin DH (2011) An economic analysis of commercial agglomeration. J Finance Econ 02:53–60
2. Tran DX, Pla F, Latorre-Carmona P et al (2017) Characterizing the relationship between land use land cover change and land surface temperature. ISPRS J Photogramm Remote Sens 124:119–132
3. Owen TW, Carlson TN, Gillies RR (1998) An assessment of satellite remotely-sensed land cover parameters in quantitatively describing the climatic effect of urbanization. Int J Remote Sens 19(9):1663–1681
4. Zhou W, Qian Y, Li X et al (2014) Relationships between land cover and the surface urban heat island: seasonal variability and effects of spatial and thematic resolution of land cover data on predicting land surface temperatures. Landscape Ecol 29(1):153–167
5. Guo G, Wu Z, Xiao R et al (2015) Impacts of urban biophysical composition on land surface temperature in urban heat island clusters. Landsc Urban Plan 135:1–10

Chapter 6
Conclusion and Prospect

This chapter aims to summarize the main conclusions and innovations of this research, as well as to outline valuable areas for future exploration in this field. In the conclusion section, we found that the intensity of the urban heat island exhibited significant spatial differentiation within the city, which was influenced by urban form, water bodies, major urban roads, and key industrial parks. Among numerous urban morphology indicators, we identified those that had a more significant impact on heat island intensity. These indicators directly or indirectly affect heat island intensity through mechanisms such as convective heat dissipation and radiative heat exchange. These influencing mechanisms inspire us to derive urban planning strategies for mitigating the thermal environment in different types of areas.

6.1 Research Conclusion

6.1.1 Spatial Pattern of Urban Heat Islands and Different Heat Island Intensity Zones

The heat island effect in Wuhan demonstrates a widespread distribution across the main urban area, characterized by a "U"-shaped ring structure with high-intensity zones on the inner and outer edges and a collapse in the middle. The pattern also features a high-temperature axis with multi-axis extensions and a mix of local high-temperature aggregation and low-temperature dispersion. Extreme and strong heat island zones are concentrated in the first ring, expanding along roadways in the second ring and aggregating around the third ring. Conversely, relatively strong and weak heat island zones are notably associated with riversides and lakes, generally located at the edges of built-up areas. While weak and extreme heat island zones are small and scattered, relatively strong heat island zones are more extensive. The spatial

Y. Lu, *Urban Form and Urban Heat Island*, Urban Sustainability,
https://doi.org/10.1007/978-981-96-3541-2_6

distribution of these zones shows a gradient decrease in intensity, with adjacent levels nested in a gradient pattern.

6.1.2 Core Urban Form Indicators Affecting Heat Island Intensity

By analyzing the urban form characteristics of typical heat island areas and comparing the urban form indicators across different heat island intensity zones, ten core indicators with a significant impact on heat island intensity have been identified. These indicators are: the proportion of industrial land, the proportion of commercial land, population density, the proportion of soil cover, NDVI, the proportion of vegetation cover, the green space rate, building density, the proportion of construction land, and GSVF. Each of these indicators has passed credibility tests, confirming their strong association with heat island intensity.

However, the current regulatory management of urban form indicators is inadequate in mitigating the heat island effect. This shortcoming arises primarily because existing regulations focus on land use functions and construction intensity while neglecting vegetation coverage levels. Although the green space ratio is a key regulatory indicator, it only accounts for large-scale green areas and overlooks smaller-scale green spaces such as building-attached greenery, rooftop gardens, and road vegetation. Research indicates that the proportion of vegetation coverage and NDVI significantly impact heat island intensity. Moreover, current regulations assess land use based on adherence to superior planning requirements and support for essential urban functions, without addressing the prevention of high-intensity heat islands. Therefore, future regulations should enhance the control indicator system by incorporating indicators related to ecological surface proportions and refining land use controls based on the early warning indicators identified in this book.

6.1.3 The Comprehensive Mechanism of Urban Form Affecting the Heat Island Effect

The comprehensive mechanism through which urban form affects the heat island effect involves the interplay of three key mechanisms: ① the mechanism by which urban form increases heat island intensity; ② the mechanism by which urban form reduces heat island intensity; and ③ the correction mechanism involving external environmental factors.

Specifically, urban form contributes to increased heat island intensity by enhancing artificial heat dissipation and radiation heat gain while restricting convective heat dissipation. Industrial, commercial, and residential activities elevate artificial heat dissipation. Buildings and hardened surfaces absorb solar radiation, thus increasing radiation heat gain, and the dense arrangement of buildings impedes air circulation, limiting convective heat dissipation. Conversely, urban form can mitigate heat island intensity through enhanced evapotranspiration and reduced radiation heat gain. Evaporation from ecological surfaces boosts convective heat dissipation, and a reduction in openness results in increased shadow areas, thereby diminishing radiation heat gain. Additionally, external environmental factors—such as concentrated industrial areas, major water bodies, and main city roads—play a corrective role. They impact the thermal environment of surrounding regulatory planning units, improving the accuracy of the association between urban form and heat island intensity.

6.1.4 Multidimensional Spatial Control Technology to Reduce Heat Island Intensity

The urban form control technology for reducing heat island intensity comprises three primary aspects:

- Urban form Indicator Warning and Optimization Technology: This involves managing individual blocks by implementing specific urban form optimization techniques based on a warning indicator system. When urban form indicators exceed their threshold ranges, the likelihood of developing very strong or strong heat island areas increases significantly. The technology provides targeted strategies to address these critical indicators, aiming to prevent extreme heat island conditions.
- Differentiated Guidance for Different Types of Blocks: This approach recognizes that the relationship between urban form indicators and heat island intensity varies across different block types. Based on the comprehensive mechanisms of heat island effects, blocks are categorized into five types, including Weak Heat Load, Strong Heat Dissipation, Weak Heat Load, Relatively Strong Heat Dissipation, Relatively Strong Heat Load, Relatively Strong Heat Dissipation, Relatively Strong Heat Load, Weak Heat Dissipation, Strong Heat Load, Weak Heat Dissipation. Each block type has distinct urban form indicators affecting heat island intensity. For each type, core, key, and general control indicators are established to form a differentiated urban form control system (see Table 5.12). This ensures that control measures are tailored to the specific characteristics and needs of each block type.
- Urban form Optimization for Special Function Zones: Special function zones, such as main roads, concentrated industrial areas, and large-scale water bodies, significantly influence surrounding heat island intensity. Optimization measures for these zones include increasing green coverage along main roads and industrial areas to mitigate the "superposition effect" of high. Temperature zones and reducing construction intensity in waterfront areas to extend the cooling effect

of water bodies. This approach ensures that the urban form optimization effectively addresses the unique impacts of these special function zones on heat island intensity.

6.2 Innovation

6.2.1 Construct a Core Urban Form Indicator System that Affects Heat Island Intensity

This book identifies several urban form indicators that significantly impact heat island intensity and constructs a comprehensive urban form indicator system for controlling heat island effects. The system includes ten key urban form indicators, including Proportion of Industrial Land, Proportion of Commercial Land, Density of Permanent Residents, Proportion of Soil Cover, NDVI (Normalized Difference Vegetation Index), Proportion of Vegetation Cover, Green Space Rate, Building Density, Proportion of Construction Land, GSVF (Ground Surface Vegetation Fraction), This indicator system provides a structured approach for managing urban form to mitigate heat island intensity, outlining the specific factors to monitor and adjust in urban planning and development.

6.2.2 Reveal the Comprehensive Mechanism of Urban Form on the Heat Island Effect

This book constructs a theoretical analysis framework to understand how urban form impacts heat island intensity. By quantitatively verifying and comparing the effects of various urban form indicators, the study provides a comprehensive mechanism that serves as a theoretical basis for urban form control aimed at reducing heat island intensity. The framework identifies three key mechanisms:

Strengthening Mechanism: This mechanism refers to how urban form increases heat island intensity by enhancing artificial heat dissipation and radiation heat gain, and by restricting convective heat dissipation.

Reducing Mechanism: This mechanism involves how urban form reduces heat island intensity by enhancing evapotranspiration, increasing building shadows, and thereby reducing radiation heat gain.

Correction Mechanism: This mechanism accounts for the impact of external environmental factors, such as industrial areas, water bodies, and major roads, which correct and refine the association between urban form and heat island intensity.

6.2.3 Proposed a Differentiated Urban Form Control Technology to Reduce the Intensity of the Heat Island

This book proposes a multi-dimensional spatial control technology to reduce heat island intensity, providing a practical implementation path for urban form control. The proposed control technology includes:

- Urban form Indicator Early Warning and Optimization Technology: This involves using early warning systems based on urban form indicators to identify and address potential heat island issues before they become severe.
- Urban form Differentiation Guidance Technology for Different Types of Blocks: This technology offers differentiated guidance strategies based on the unique characteristics of various block types, tailoring control measures to specific needs and conditions.
- Urban form Optimization Technology for Special Function Zones: This focuses on optimizing urban form in areas with special functions, such as main roads, industrial zones, and large water bodies, to mitigate their impact on surrounding heat island intensity.

6.3 Limitations and Prospects

6.3.1 Insufficient Research on the Impact of Urban Form on Heat Island Intensity Under Different Climatic Conditions

According to the "General Rules for Civil Building Design" (GB50352-2005), China is divided into 7 main climatic zones and 20 sub-climatic zones, reflecting significant variations in climate due to geographical latitude and terrain conditions. These climatic differences can lead to substantial variations in how urban form impacts heat island intensity. This book, focused specifically on Wuhan City, which falls within the hot summer and cold winter region, provides insights based on this particular climate. However, the applicability of these conclusions to other climatic zones remains to be verified. Further research is needed to determine how well these findings translate to regions with different climatic conditions.

6.3.2 Insufficient Research on Planning and Design Strategies to Reduce Heat Island Intensity

The urban form regulation technology proposed in this book focuses on the influence mechanisms of urban form indicators. However, it does not delve deeply into how to achieve the goals of urban form indicator regulation through specific planning and design strategies. Additionally, it lacks a discussion on how to adopt differentiated planning and design strategies for different types of blocks. Moreover, while the research addresses the reduction of high-intensity heat islands to improve summer conditions, it does not consider the need to mitigate potential adverse effects on residents during winter, such as the impact of cold air. Future work should address these aspects by developing detailed planning and design strategies tailored to different block types and ensuring that measures to reduce heat island intensity do not inadvertently worsen cold weather conditions.

The manufacturer's authorised representative in the EU is Springer Nature Customer Service Centre GmbH, Europaplatz 3, 69115 Heidelberg, Germany. If you have any concerns regarding our products, please contact ProductSafety@springernature.com

Printed and bound by CPI Group (UK) Ltd, Croydon, CR0 4YY
01/10/2026
02225941-0001